Physical Methods in Chemistry and Nano Science.
Volume 8: Applications of Analytical Methods

Physical Methods in Chemistry and Nano Science.
Volume 8: Applications of Analytical Methods

Editor

Andrew R. Barron

Contributors

Adrish Anand, Andrew R. Barron, Jagnoor Benipal, Kyle Chow,
Selin Ergulen, Angel Adrian Garces, Patrick Gilliam, Isaac Goforth,
Katharyn Hernandez, Thanh Huynh, Claire Jahnke, Patrick Kelly,
John Kim, Richard Lee, Max Liu, Sreyas Menon, Milad Najafabadi,
Michelle Nguyen, Janett Ordonez, Zoe Punske, Laura Quinn,
Pavan M. V. Raja, Evan Rebesque, Anna Reed,
Ricardo Rivera-Maldonado, Raul Rondon, Reece Rosenthal,
Steven Schara, James Siriwongsup, Sydney Stocks, Troy Tabarestani,
Tianyi (Emma) Wu, Olivia Zhang, Alice Zhu

MiDAS Green Innovations
2020

MiDAS Green Innovation, Ltd
Swansea, SA1 8RD, UK

www.midasgreeninnovation.com

Dedication

To the management and staff of the companies that were founded on my research.

"He who dies with the most toys.... Is still dead"

Marillion (2007)

Contents

Acknowledgements vii

Preface ix

Chapter 1: Industry Overview 11

Chapter 2: Environmental Conservation and Safety 21

Chapter 3: Chemical Industry 39

Chapter 4: Oil and Gas Industry 49

Chapter 5: Pharmaceutical Industry 65

Chapter 6: Automotive Industry 87

Chapter 7: Semiconductors and Nanotechnology 107

Chapter 8: Forensic Science 117

Chapter 9: Detection of Explosives 129

Chapter 10: Outer Space 141

Acknowledgements

I would like to thank all the contributors to this Volume, for their interest in creating a user-friendly text for their peers.

The co-authors and I would like to acknowledge the input and collaboration of the following for specific Chapters:
- Chapter 1 - Zeke Ramirez and Andy Yuwen
- Chapter 2 - Sammy Hamdi
- Chapter 3 - Ben Baldazo, Alexander Hawley, and Sonia Rao
- Chapter 4 - John Komoll, Leo Onor, and Mekedlawit Setegne
- Chapter 6 - Hyoungjoo Kim and Matthew Perez
- Chapter 7 - Alejandro Akerlundh.

The myriad collaborators of whom I have had the pleasure of learning from over the years are to be thanked for bringing new characterization methods to my research - you know who you are.

Last, but not least, I would like to thank my wife, Merrie, for putting up with me during the Editing of this book while 'social distancing' during the COVID-19 pandemic.

viii

Preface

This Series intended as a survey of research techniques used in modern chemistry, materials science, and nanoscience. The topics are grouped into volumes, not be method *per se*, but with regard to the type of information that can be obtained. Thus, the Volumes are ordered as follows:

- Elemental composition.
- Physical and thermal analysis.
- Chromatography
- Chemical speciation.
- Molecular and solid state structure.
- Surface morphology and structure at the nanoscale.
- Device performance.
- Applications of analytical methods

As a consequence of this organization methods can be found in different Volumes. For example, X-ray photoelectron spectroscopy is included under Elemental Composition (Volume 1) with regard to its use for determining the chemical composition, while it is included under Chemical Speciation (Volume 3) with regard to determining the identity of component chemical moieties.

The goal was to create simple to understand explanations of methods that allow the reader to gain the knowledge to correctly apply a technique or interpret data. As a consequence, the topics in this book have been developed in partnership with undergraduate and postgraduate students at Rice University over a 7-year period, and because of this there is some variation in depth and focus given to each topic. I make no apology for this diversity.

Chapter 1: Industry Overview

John Kim and Richard Lee, Pavan M. V. Raja and Andrew R. Barron

Introduction

Analytical chemistry is the study of the separation, identification, and quantification of the chemical composition of materials. Analytical techniques play crucial roles in industry to confirm purity of products, conclusively identify and analyze properties of chemical compounds. Modern analytical systems allow accurate and automated qualitative or quantitative chemical analysis for rapid and reproducible data. Examples of techniques include:

- chromatography used to separate compounds for identification or purification,
- spectroscopy analyzes the interaction between compounds and electromagnetic radiation to determine properties such as chemical structure.

Analytical chemistry finds application in practically every industry as chemicals are ubiquitous and require detailed, precise analysis to ensure safety and quality. The most common and mass-produced chemicals are summarized in Table 11.1.

Chemical or Material	Global production in 2016 (million metric tons/year)	Applications
Steel	1621	Construction, automobiles, metal goods
Sulfuric acid (H_2SO_4)	264	Fertilizers, paper, car batteries
Ethylene ($H_2C=CH_2$)	146	Polymers, chemical building block
Propylene ($H_2C=CHMe$)	99	Polymers, chemical building block
Sodium hydroxide (NaOH)	72	Chemical production, paper, surfactants
Hydrogen (H_2)	50	Fertilizers (ammonia), oil refining

Table 1.1: A list of globally mass-produced chemicals and materials.

Summary of analytical techniques

Before a chemist decides what technique to use, they will need to critically question what kind of information they need. The flowchart in Figure 1.1 summarizes some basic questions to ask about the material to be analyzed, and what technique to apply will depend on the answers. Some techniques will only work on solids or only on gases; others will require solubility in a certain solvent or a specific kind of chemical composition. The toxicity of the material must also be determined to ensure the proper safety and disposal precautions. Finally, for industry, practical considerations such as cost and time are essential, as sometimes there is such a thing as too accurate if the cost and time of analysis is not feasible for a company, while other times purity is so important (for example, in the semiconductor industry) that a company will invest significant resources to ensure quality.

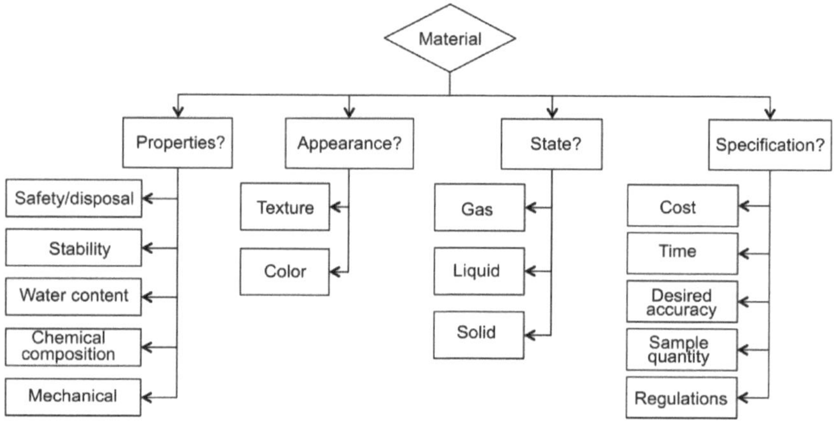

Figure 1.1: A flowchart detailing questions to ask in analyzing a particular material sample.

There is a wide range of analytical techniques commonly used in industry and academia (e.g., Table 1.2). The leftmost column lists broad categories of techniques, and the middle column gives a few examples within each category. For example, spectroscopy is an extremely useful set of techniques, including NMR and IR to investigate chemical composition, and spectrophotometry to determine unknown concentrations of samples. Other techniques have important applications as well, filling up the chemist's toolkit to conduct all kinds of analysis.

Category	Specific Techniques	Application
Spectroscopy	NMR, IR, spectrophotometry	Composition, concentration
Gravimetry	Thermogravimetric	Mass
Calorimetry	Differential scanning, bomb	Exothermicity
Titration	Direct titration, back-titration	Concentration
Electrochemistry	Electrolysis, electrogravimetry	Purification, mass
Chromatography	Column, thin layer, HPLC	Purification, concentration

Table 1.2: Specific analytical techniques commonly used in industry.

Specific industrial sectors

Energy

The oil and gas industry is generally partitioned into four distinct sectors:
- upstream,
- oil field services (OFS),
- midstream,
- downstream.

Analytical chemists play unique roles and face idiosyncratic challenges in each of these sectors.

Upstream

Upstream oil and gas encompass the exploration and production (E&P) of crude oil and natural gas. Leaders in the upstream sector include ExxonMobil, Shell, Saudi Aramco, BP, and many other large, integrated corporations.

Oil field services

Oil field service (OFS) companies, such as Schlumberger, Halliburton, and Baker Hughes, are contractors in the E&P sector who provide services at the point of extraction. As a result, many analytical chemists work in diverse roles in the OFS sector. For instance, they can analyze the composition of the extracted oil, or analyze wastewater quality to prevent negative environmental externalities. The extraction process can also vary widely geographically. For instance, Canadian petroleum deposits, which are predominantly centered in Alberta, are often oil sands. These consist of water, sand, clay, and bitumen, which is an extremely viscous and dense form of petroleum. As a result, oil extraction in Canada involves very costly and extensive steam processing.

Analytical chemists work alongside process engineers to ensure that the system is properly controlled, minimizing the risk of costly inaccuracies.

Midstream

The midstream sector oversees the transport of resources from the point of production to the downstream refinery. This transport generally occurs by pipeline, rail, truck, or oil tanker. Prominent midstream companies include Kinder Morgan, Enterprise Products Partners, and Enbridge. Because the midstream oil and gas sector is concerned with infrastructure, analytical chemists in the sector are primarily concerned with maintenance and risk prevention. For instance, analytical chemists are involved in the design of oil pipelines to prevent rusting.

Downstream

The downstream oil and gas sector encompass the refining of crude oil and natural gas to produce industrially useful materials such as liquefied natural gas (LNG), gasoline, lubricants, pharmaceuticals, plastics, asphalt, fertilizers, and many others. Analytical chemists in the downstream sector must first determine the composition of the input crude oil and natural gas, while also performing quality assurance (QA) to ensure the output products meet customer specifications and comply with government regulations.

Chemicals

The chemicals industry is broad and includes various other sub-industries, such as paper/pulp, downstream oil and gas, catalyst production, lab reagents, adhesives, etc. As a result, the role of analytical chemists and the techniques they utilize vary immensely across the industry. However, certain roles are staples of any production process, including the determination of input feed composition, quality assurance of customer product specifications and waste categorization for regulatory compliance.

Automotive and aerospace

In the automotive and aerospace industries, chemical analysis is implemented to ensure quality/compliance and safety of the products and to address environmental concerns (Figure 1.2). In the automotive industry, for example, end-of-life vehicle directive was implemented to reduce hazardous substances in vehicles to minimize their release to the environment, facilitate proper dismantling of the vehicle to allow components and materials to be reused, recycled and/or recovered. Also, the Clean Air Act, led by the Environmental Protection Agency (EPA), regulates fuels and fuel additives used in automotive to ensure that they are not harmful to the environment. Regarding industrial coatings in automotive or aerospace, chemical analysis ensures sustainable and non-polluting manufacturing processes as well as the manufacturing of products that do not release harmful compounds into the air or water.

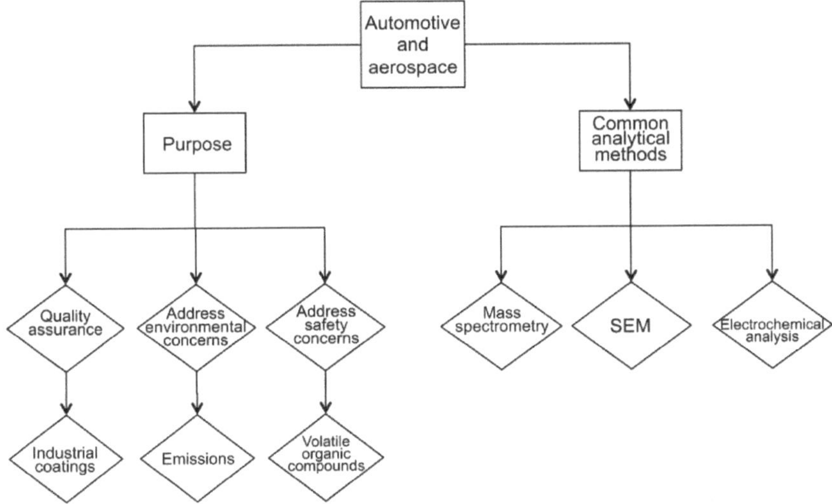

Figure 1.2: Chemical analysis in the automotive and aerospace industry.

For quality assurance, ASTM International's industrial chemical standards allow the testing and evaluation of the physical and chemical properties of substances used or produced in industrial applications. These industrial chemical standards allow producers and users of such chemicals in their proper fabrication and treatment processes to ensure quality towards safe production and utilization. Some methods of chemical analysis employed in the automotive industry include mass spectrometry and electrochemical analysis. In the aerospace industry, chemical analysis of persistent contrails is done to regulate aircraft emissions that may potentially harm the environment and humans. Also, various methods of chemical analysis are used in aerospace laboratories, including the use of scanning electron microscopy, Fourier transform infrared spectroscopy, and non-destructive testing (NDT).

Food, pharmaceuticals, and healthcare

Analytical chemistry is employed extensively in the pharmaceuticals, food, and healthcare industries (Table 1.3). In pharmaceuticals, the analytical investigation of drugs, including bulk drug materials, intermediates, drug products, and drug formulations, is imperative to the industry as quality assurance of the drugs are important. Chemical analysis allows the understanding of the physical and chemical stability of drugs, and quantitation and identification of impurities, which is essential to evaluating the toxicity pro les of these impurities. Some common analytical techniques implemented in pharmaceuticals include titrimetric, chromatographic, spectroscopic, electrophoretic, electrochemical, and colorimetric methods.

Type of industry	Analytical methods
Pharmaceuticals	Titrimetry, chromatography, spectroscopy, electrochemistry, colorimetry
Food	Gas chromatography, high performance liquid chromatography, mass spectroscopy
Healthcare	Colorimetry, immunoassay

Table 1.3: Common analytical methods employed in the pharmaceutical, food, and healthcare industry.

Food products need to meet government regulations and recommendations set by the US Food and Drug Administration (FDA) and the quality and safety of the products need to be concerned. Chemical analysis of food products can also help provide information about the molecular relationships between food ingredients, nutrient composition, cellular function, and organism metabolism. Gas chromatography, high performance liquid chromatography, and mass spectroscopy are commonly used for food analysis. In healthcare, chemical analytical methods are widely used, such as blood glucose tests for diabetes that utilize colorimetry, urine and pregnancy tests that use immunoassays. Chemical analysis in the healthcare industry is also important in order to ensure the safety and quality of the materials used. For example, medical air/oxygen is used to aid in long-term life support as in intensive care units, critical care units, and neonatal intensive care units, therefore, quality assurance is important.

Environment, safety, and toxicology

As public awareness for environmental protection has increased, the need for analytical chemists in the environmental and safety field has risen. The goal of analytical chemistry in these fields is to assure compliance with environmental and other regulations, as determined by specific agencies. The Environmental Protection Agency (EPA) is the regulatory unit in the United States, and they are authorized to write regulations that explain the technical, operational, and legal details necessary to implement laws. A few of the sectors that the EPA writes regulations for include agriculture, the automotive industry, and oil and gas. In all areas of environmental protection, the analytical chemist is intimately involved because almost every control or regulation relates to permissible concentration levels of substances, which may be present in discharges, effluents, etc. Their analysis is therefore a prerequisite to the effective implementation of those controls and regulations (Figure 1.3).

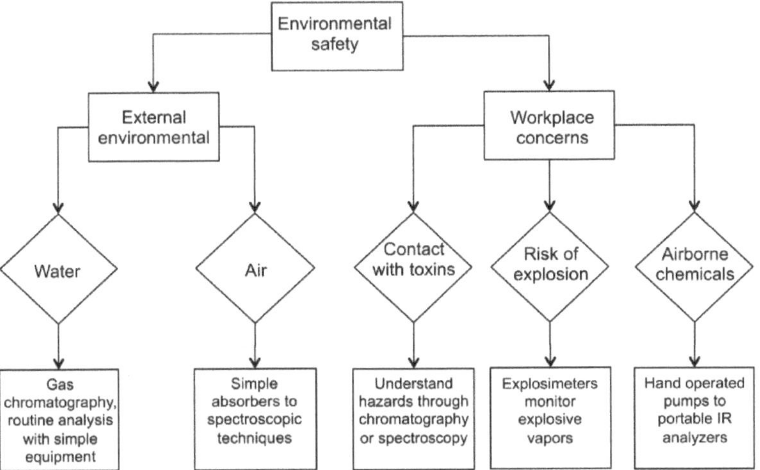

Figure 1.3: A flowchart detailing the analytical processes related to each aspect of environmental safety.

The regulations of the environment and safety field can be applied to two different categories:

- the external environment,
- the workplace.

The external environment concerns both air and water quality (Figure 1.3), and both have different analytical procedures. Analytical techniques used to monitor air quality vary greatly and range from simple absorbers to sophisticated and spectroscopic techniques, and automated procedures for measuring many pollutants have been developed. For water, routine analysis is relatively uncomplicated, only needing simple equipment and minimal expertise (Figure 1.3). However, more complex procedures such as gas chromatography are required to deal with certain pollutants.

In the workplace, there are three main areas of risk to workers from the use of chemicals:

- contact with toxins,
- risk of explosion,
- inhalation of airborne chemicals and dusts.

Direct contact with toxic, carcinogenic or corrosive materials can be negated by understanding the hazards associated with a chemical, which an analytical chemist will do with techniques like chromatography or spectroscopy. Explosimeters can be easily operated by non-analytical chemists to monitor explosive vapors. Analytical techniques used to compare airborne vapors to their Threshold Limit Value (TLV) range from hand operated pumps to portable infra-red analyzers.

Semiconductors, electronics, and nanotechnology

As nanotechnology becomes more common and impactful, its connection to analytical chemistry becomes more important. Nanomaterials can be characterized and/or determined in a variety of samples, such as cosmetics, agrifoods, and clinical and environmental matrices. The valuable biochemical information of analytical chemistry provides a significant support to the integral development of nanoscience and nanotechnology. On the other hand, nanotechnology is also used to analyze other compounds. Nanomaterials can play different roles (sorbents, stationary and pseudo-stationary phases, inert and active supports, fluorophore probes, and electric conductors) to improve the existing and develop new analytical processes, aiming to exploit the unique physicochemical properties of the nano-matter, thereby enhancing the analytical properties, which are indicators of the quality of the information provided.

Analytical chemistry was traditionally used to determine atom concentrations through techniques like spectroscopy, but now it is also used to analyze electronic solid-state materials, which includes semiconductors (Figure 1.4).

Because semiconductors are electronic by nature, analytical techniques such as secondary ion mass spectrometry lead to discovering chemical information with the electrical properties desired. By pro- viding this chemical information, analytical chemistry plays a vital role in the development of semiconductor materials and processing schemes.

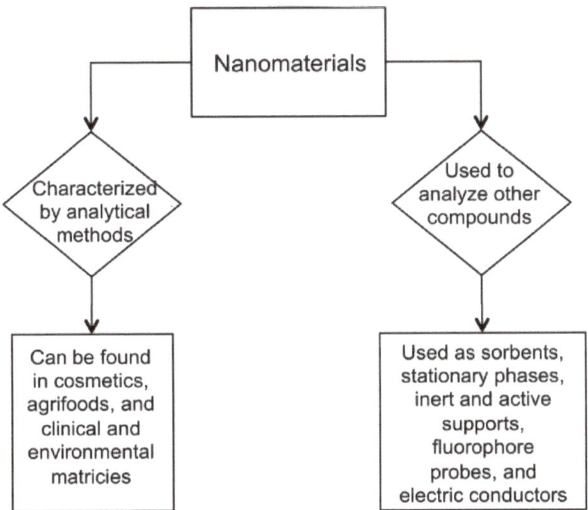

Figure 1.4: A flowchart detailing how nanomaterials are characterized through analytical processes but are often used to characterize other materials.

Conclusions and future directions

Analytical chemistry has played a critical role especially as environmental and safety concerns have increased in recent years, so it can be expected that faster and more accurate methods will be developed that can be applied not only with specialized equipment but also in the field with handheld equip- ment. This dramatic improvement is made possible with the computers found in everyday life, and the techniques currently used in the industries addressed in this chapter and others will only improve in reliability over time.

Bibliography

ASTM International, https://www.astm.org/Standards/industrial-chemical-standards.html.

P. W. Bohn and T. D. Harris, Analytical chemistry and semiconductor materials. *Anal. Chem.*, 1990, **62**, 767A.

Federal Gasoline Regulations, https://www.epa.gov/gasoline-standards/federal-gasoline-regulations. .

Natural Resources Canada, http://www.nrcan.gc.ca/energy/oil-sands/5853.

Á. I. López-Lorente and M. Valcárcel, The third way in analytical nanoscience and nanotechnology: Involvement of nanotools and nanoanalytes in the same analytical process. *Trends Anal. Chem.*, 2016, **75**, 1.

M. R. Siddiquia, Z. A. AlOthman, and N. Rahman, Analytical techniques in pharmaceutical analysis: A review. *Arab. J. Chem.*, 2017, **10**, S1409.

Steel Statistical Yearbook 2016, https://www.worldsteel.org/en/dam/jcr:37ad1117-fefc-4df3-b84f-6295478ae460/Steel+S.

US Environmental Protection Agency Regulations, https://www.epa.gov/laws-regulations/regulations.

Chapter 2: Environmental Conservation and Safety

Isaac Goforth, Max Liu, James Siriwongsup, Pavan M. V. Raja
and Andrew R. Barron

Introduction

As economies around the world continue their ceaseless march towards growth and expansion, environmental pollution has become an issue on a global scale. As such, the development of effective analytical techniques to monitor, assess, and evaluate the air, water, and soil is an integral part in keeping pollution levels in check. Not only do environmental chemists measure such levels quantitatively, but they also must qualitatively gauge their results to understand how chemicals and the human activity associated with its production affect the environment. Because the environment and human safety are inextricably linked (many chemicals that are known to pollute the environment also adversely affect your health) it is imperative that environmental chemists communicate their findings to industries responsible for releasing pollutants in order to address the issues on regulatory terms.

Overview of pollutants

The primary pollutants in air include carbon monoxide, oxides of nitrogen and sulfur, particulates, and volatile organic compounds (VOCs) in the atmosphere. While all of these pollutants can be naturally produced, their levels in the atmosphere are exacerbated by numerous manmade (anthropogenic) sources. A major source of carbon monoxide is the incomplete combustion of fossil fuels, a source, which is also responsible for releasing many of the greenhouse gases commonly attributed to global warming.

Nitrogen oxides (NO_x) and sulfur oxides (SO_x) are primarily produced from the burning of fuels by power plants and other industrial facilities. Nitrogen dioxide (NO_2), which is an oxide of great environmental and health concern, forms from the emissions of vehicles, power plants, and off-road equipment. Sulfur dioxide (SO_2), on the other hand, is produced from the extraction of metal from ore and specific industries that directly utilize or mass-produce sulfur-containing compounds such as those performing the contact process, which is a method of producing sulfuric acid in high concentrations needed for various industrial processes.

Particulate matter refers to a mixture of solid particles and liquid droplets found in the air. They range from submicroscopic particles emitted from combustion of organic compounds and metals to those that can be seen with the naked eye such as dirt, soot, or smoke. Because particulate matter is a wide-ranging pollutant that encompasses hundreds of compounds and materials, their sources are general and can be attributed to any industry or process that releases waste products into the air. In particular, diesel exhaust from automobiles release soot and metal particles into the air producing smog. Figure 2.1 shows a graphic representation of global PM2.5 pollution. PM2.5 refers to atmospheric particulate matter (PM) that have a diameter of less than 2.5 µm. Interestingly, heavy pollution is concentrated in North Africa, the Middle East, South Asia, and China, where a significant portion of the population burn wood or other organics for cooking and heating.

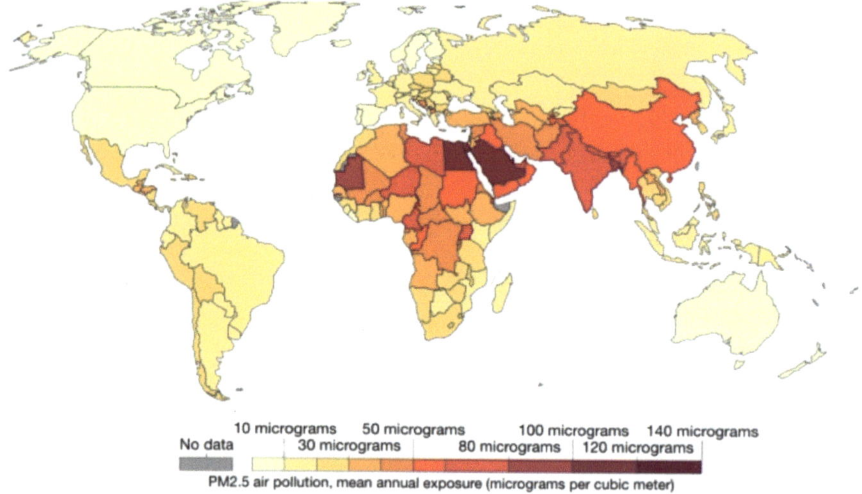

Figure 2.1: Global population weighted exposure to ambient PM2.5 pollution (µg/m₃). Adapted from H. Ritchie and M. Roser, *Air Pollution*, https://our-worldindata.org/air-pollution/.

The release of volatile organic compounds (VOCs) is often ascribed to sources that are associated with fumes; namely, industries that produce paint, cleaning supplies, and solvents as well as processes that produce unburned gasoline and fuels.

The primary pollutants found in wastewater include heavy metals, pesticides, dioxins (Figure 2.2), polychlorinated biphenyls (PCBs, Figure 2.3), nitrates, and phosphates. Among the heavy metals, As, Cd, Pb, Cr, Cu, Hg, and Ni are of major concern, mainly due to their presence at relatively high concentrations in drinking water and their known detrimental effects on human health. These heavy metals may leach into our drinking water supply through industrial processes such as mining, mineral processing, and ore production. Products such as batteries and paints may also contain heavy metals and can contaminate the water if improperly disposed.

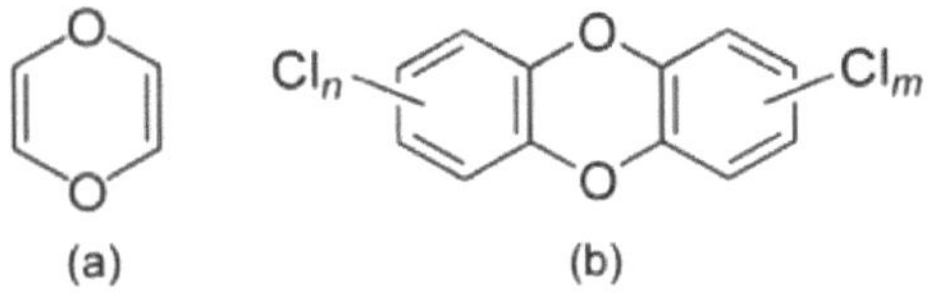

Figure 2.2: Structure of (a) 1,4-dioxin, and (b) the chlorinated dioxins polychlorinated dibenzodioxins.

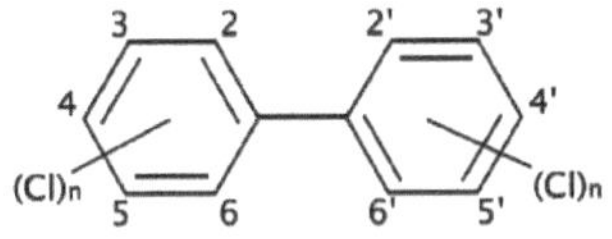

Figure 2.3: General structure of polychlorinated biphenyls (PCBs).

As an example, to illustrate the presence of heavy metal poisoning in urban communities, Figure 2.4 depicts the prevalence of lead poisoning in New York City for children under 6 years old from 2006 - 2015. Most neighborhoods have a prevalence of under 5%, but some parts of the city suffer from a prevalence as high as 20%. Since the 1970s, New York City has dramatically reduced the number of children exposed to lead\ but hasn't met its longstanding goal of eliminating childhood poisoning, leaving some children vulnerable. Along with lead-based paint in old housing, the city's children can face poisoning hazards from consumer products, soil, and water.

Pesticides, along with organic compounds containing nitrates and phosphates, can be released in the water from agriculture and the use of fertilizers. Farms are a major nonpoint source of runoff, and rainwater and irrigation can potentially drain fertilizers and pesticides into bodies of water. Polychlorinated biphenyls are a group of chlorinated hydrocarbons and are used in numerous industrial and commercial applications including electric equipment, plasticizers, pigments, dyes, and rubber products. Although manufacturing of PCBs was banned in 1979, PCBs may be present in products and materials produced before the ban and released into the environment. Finally, dioxins are a group of persistent environmental pollutants and are highly toxic. While the term dioxin refers to a specific compound 2,3,7,8-tetrachlorodibenzo-p-dioxin (TCDD, Figure 2.5), the name is often used to describe a larger family of structurally similar derivatives (Figure 2.2). They can be produced specifically through the production and incineration of chlorinated organic compounds and plastics.

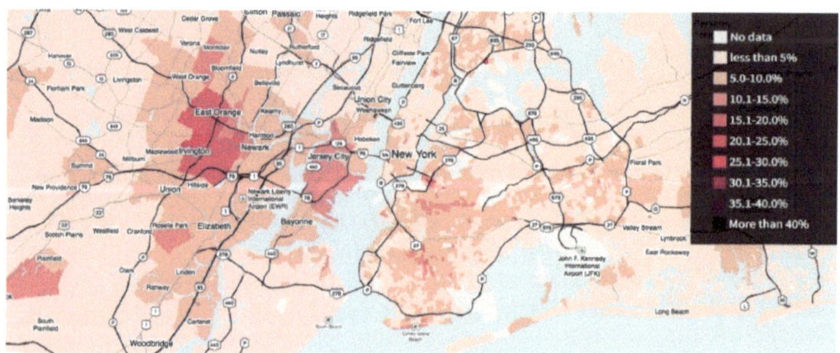

Figure 2.4: Prevalence of childhood lead poisoning in New York City from 2006 - 2015. Adapted from M. B. Pell and J. Schneyer, *The thousands of U.S. locales where lead poisoning is worse than in Flint*. Reuters. December 19, 2016. https://www.reuters.com/investigates/special-report/usa-lead-testing/.

Figure 2.5: Structure of 2,3,7,8-tetrachlorodibenzo-p-dioxin.

Many pollutants of the ground overlap with the primary pollutants of water simply due to the fact that water contaminants can easily leach into the surrounding soil. However, some pollutants that are primarily associated with the soil include antibiotics, polycyclic aromatic hydrocarbons (PAHs), polymers, and nanomaterials. PAHs are insoluble in water, which limits their mobility in aquatic environments. However, they can be found in the ground from leached chemicals that occur naturally in coal, crude oil, and gasoline, especially upon the combustion of these chemicals. PAHs can also bind to or form small particles in the air resulting in an additional source of air pollution.

The pollution from antibiotics refers to the large-scale release of antibiotics into the environment. Not only do antibiotics deteriorate environmental quality, but they also result in the appearance and rapid spread of resistant bacteria in the environment. Therefore, unique from other sources of pollution, antibiotic pollution is of increasing concern as it poses additional public health risks associated with bacterial resistance.

The pollution of polymers manifests itself most commonly in the form of plastics. Because plastics are slow to degrade, they often rapidly accumulate in both the land and the oceans. Due to such a wide-reaching range, living organisms are often victims of plastic pollution, and entire ecosystems have been severely affected when organisms directly ingest these materials, or are exposed to them. Lastly, a relatively overlooked threat to the environment is that of nano pollution. Nanoparticles are some of the smallest macromolecular pollutants and are being expelled in increasingly greater amounts from the booming nanotechnology industries.

Table 2.1 lists all of the aforementioned pollutants alongside their primary anthropogenic source, main health effects, and comments on their industrial production. Figure 2.6 arranges selected pollutants in a Venn diagram as a way to conceptualize the prevalence of pollutants in the environment. flowing to their small size (on the atomic level), heavy metals are an example of a pollutant that can be present in all three domains: air, water, and ground.

Pollutant	Anthropogenic source	Health effects	Production
Carbon monoxide (CO)	Combustion of fossil fuels, coal power plants, waste incinerators	CO poisoning, heart disease	2.6 billion tons (global), anthropogenic (60%), deforestation (40%)
Nitrogen oxides (NO_x)	Power stations, combustion engines, heavy industry	Respiratory diseases, lung infections	14.1 million tons (US, 2011)
Sulfur dioxide (SO_2)	Coal/oil burning, smelting of metal ores, contact process. 99% of all SO_2 emissions come from man-made sources	Respiratory diseases, lung infections	6.3 million tons (US, 2011)
Continued on next page			

Pollutant	Anthropogenic source	Health effects	Production
Particulates	Dust from industrial activity, soot from incomplete combustion	Lung diseases, blood clots/heart attacks, in- creased susceptibility to viruses and bacteria	EPA standard: annual 12.0 µg/m^3 (primary), 15.0 µg/m^3 (secondary)
Volatile organic compounds (VOCs)	Combustion engines, paint solvents, varnishes	Liver, kidney, CNS damage, eye, nose, throat infections, cancer	12.3 million tons (US, 2011)
CFCs	Refrigerants, propellants, cleaning solvents	Nausea, diarrhea, irritation of digestive tract	1 million tons (global, in 1970s), causes ozone depletion; production phased out by 1996
Heavy metals	Mining, mineral processing, production and smelting of ores, cement production, waste incineration	Neurotoxicity and brain dysfunction, metal poisoning, kidney damage, various cancers (Figure 2.4)	
Pesticides	Agriculture/run-off, agriculture accounts for 80% of pesticide use	Skin/eye irritation, nerve damage	5.2 billion tons (global), 1 billion tons (US)
Dioxins	Production and incineration of chlorinated organic compounds, plastic production	Skin lesions, impairment of the immune system, liver damage	Banned in the US in 1979, but still may still be present in the environment
Polychlorinated biphenyls (PCBs, Figure 2.2a)	Electric transformers and capacitors	Acne, rashes, liver damage, stomach damage	1.2 million tons between 1930 and 1975 (US), banned in 1976 (US)
Nitrates and phosphates	Fertilizes, agriculture	Nausea, diarrhea, vomiting, kidney/heart damage	
Antibiotics	Urban environments, agriculture	Increasingly resistant bacteria	
Continued on next page			

Pollutant	Anthropogenic source	Health effects	Production
Polycyclic aromatic hydrocarbons (PAHs)	Combustion of coal, oil, wood, residential heating, coke and Al production	Irritation of eyes and breathing passages, blood, liver abnormalities	300 - 1300 metric tons released annually (US)
Polymers (plastics)	Plastic industry land fills	Disrupted endocrine system	299 million tons of plastics produced (globally)
Nanomaterials	Manufacturing	Largely unknown, but probably similar to effects of heavy metals	

Table 2.1: Overview of major environmental pollutants.

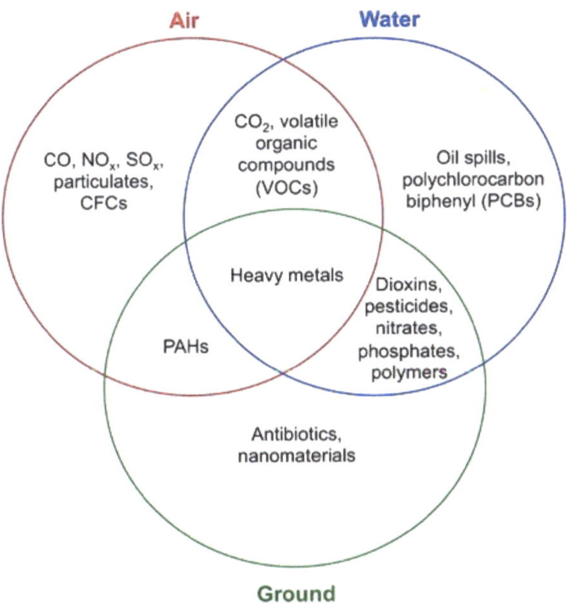

Figure 2.6: Venn diagram of major environmental pollutants.

Regulations protecting the environment

The momentous advances the chemical industries have made have not come without their deleterious environmental effects (both intentional and unintentional alike. But a multipronged approach starting in the 1970s) driven by environmentalists and their allies in the private and public sectors, has made tremendous progress in limiting the adverse environmental impacts of many industrial processes. In fact, the year 1970 marked a major win for environmental protection when the US Environmental Protection Agency (EPA) was founded, and the Clean Air Act was passed. From reducing smog and toxic chemicals emitted from cars to disposing of industrial wastes in places other than our oceans and rivers to the elimination of harmful pesticides and refrigerants in favor of more benign replacements, much progress has been made in understanding the fate and transport of chemicals in the major environmental sinks and adjusting as needed to avoid major and gradual environmental catastrophes alike.

It was not an easy sell to chemical industries when their bottom line was invariably affected, but the intense public backlash to the widespread chemical waste in our environment made industry realize it was in their interest to work in improving their use, handling, disposal of critical chemicals as well as improving remediation and detection techniques using analytical tools such as chromatography and mass spectrometry. Industry and government led this change alike, the former providing crucial guidelines and industry standards and practices and the latter setting stricter rules and regulations to maintain environmental integrity.

The primary obstacle to the collaboration between the public and private sectors to protect the environment from harmful chemicals produced in industry is balancing the needs of industry to maintain proprietary information for competitive advantage and the needs of the public to be cognizant of potentially harmful chemicals. The major challenge moving forward is more intensive and accurate mathematical modeling and understanding of chemicals in the environment and sampling of our air, water, and soil that will require the collaboration of industrialists, research chemists, and engineers alike.

This life-cycle analysis of industrial chemicals and system approach the environmental modeling is now very much emphasized in industry and academia alike, and collaboration of both sectors moving forward is critical to environmental protection.

Air

The Clean Air Act (1970) is the major, comprehensive federal law in the US to regulate harmful air emissions and pollutants. This law allows the EPA to establish Ambient Air Quality Standards to protect public health by directing states to develop actionable plans to set certain emission standards and achieve emission goals. Additionally, this act allows the EPA to set pollution limits of air toxics like benzene, lead, and perchloroethylene from large industrial facilities. This is done through two phases, the first of which is a technology standard set by what is maximally achievable with current technology based on levels being currently achieved in low-emitting facilities in industry. The second phase is a risk assessment phase where the EPA evaluates whether the margin of safety of the standards is sufficient to protect the environment and manage health risks. The EPA can change these standards in light of new pollution controls and prevention technology. Under this law, the EPA also sets maximum ambient limits for six criteria pollutants ozone (O_3), particulate matter, carbon monoxide (CO), nitrogen dioxide (NO_2), sulfur dioxide (SO_2), and lead based on the technology standard (Table 2.3).

Pollutant	Level	Emitting time period
Carbon monoxide (CO)	9 ppm, 35 ppm	8 hours, 1 hour
Lead (Pb)	0.15 μg/m^3	3 months
Nitrogen dioxide (NO_2)	53 ppb, 100 ppb	1 year, 1 hour
Ozone (O_3)	70 ppb, 125 ppb	8 hours, 1 hour
Sulfur dioxide (SO_2)	75 ppb	1 hour
Particulate matter (10 μm)	150 μg/m^3	24 hours
Particulate matter (2.5 μm)	12 μg/m^3, 35 μg/m^3	1 year, 24 hours

Table 2.3: National ambient air quality standards. Data obtained from EPA. *Criteria Air Pollutants*: **NAAQS Table, https://www.epa.gov/criteria-air-pollutants/naaqs-table.**

Rather than directly set emission limits for the criteria pollutants, the EPA directs states to maintain ambient pollutant standards by issuing permits to major facility emitters. Greenhouse gases like carbon dioxide (CO_2) and methane (CH_4) were not covered originally because they were not thought to directly endanger public health, though in 2007 the Supreme Court ruled that under the Clean Air Act, the EPA has the obligation and authority to do so if they in fact endanger public welfare. Under the Obama administration, the EPA proposed the Clean Power Plan for existing power plants to reach CO_2 emission rates state by state and carbon pollution standards for new power plants at 1000-1100 lb/MWh, although many of these rules are currently back-logged by challenges in the court system.

Surface water

The Clean Water Act (1948) and Safe Drinking Water Act (1974) allowed the EPA to implement pollution control programs, primarily setting wastewater discharge standards for industry and municipalities and water quality standards in regard to pollutants/contaminants in surface and drinking waters. These acts authorize the EPA to establish guidelines for testing procedures used by industries and municipalities for gathering data and ensuring compliance with set effluent or wastewater standards. These guidelines are a compilation of methods published by the consensus standards organization and others in the environmental community. Organizations who want to utilize alternative methods can do so only with EPA validation of and rulemaking on the proposal, termed the alternative test procedures (ATP). This allows professionals in the industry to utilize emerging technologies and novel analytical techniques that the government is slow to pick up on to ensure more accurate monitoring of chemicals for compliance.

Permits need to be obtained for industrial facilities disposing of chemical waste and discharging wastewater. However, of particular importance in the chemical industry is incidental or accidental chemical discharge into our waterways, particularly in the oil and gas industry. Shale gas extraction produces large volumes of wastewater from hydraulic fracturing that can contain high concentrations of dissolved solids (salts), heavy metals, and other pollutants like lubricants and antibacterial chemicals used in drilling and completion of wells. This produced water is tightly regulated by permits allowing injection into deep wells so that it cannot enter the water table. Additionally, the EPA's Oil Pollution Prevention regulations further require owners and operators of oil and gas facilities to make and implement plans to prevent discharges. The EPA will periodically conduct inspections on oil and gas processing facilities based on such factors as proximity to drinking water intakes or environmentally sensitive areas like wetlands and places with endangered species, or the condition of infrastructure that can leak like tanks or piping.

The Toxic Substances Control Act (1976) gives the EPA authority to regulate the production, use, and exposure of certain toxic chemicals by mandating facilities to keep extensive records and monitoring of designated toxic chemicals in their wastewater and their eventual fate in the environment. Until

2016, the EPA had to determine that a chemical presented an unreasonable risk given its use to obtain proprietary information about it, in contrast to other places like the EU where industry is tasked with proving chemicals are safe. In 2016, the Chemical Safety Act for the 21st Century instituted mandatory safety review of chemicals and greater transparency on the model of the EU regulations. Before, many state regulations had gone far beyond federal standards due to acute human health risks of these chemicals, leading to competitive advantage issues in industry. Some chemicals used in industry of particular import that fall under this act for their harmful effects to humans and the environment are polychlorinated biphenyls (PCBs), heavy metals, asbestos and radon.

Soil and groundwater

The final major environmental sink that is regulated by the EPA is the ground. Two major concerns with soil and groundwater contamination are nutrient and pesticide runoff from farms as well as leakage from underground storage tanks, e.g., nuclear waste repositories or gasoline/chemical storage tanks. The Department of Energy is responsible for maintaining nuclear waste sites under the Nuclear Waste Policy Act (1982). The EPA is then tasked with protecting the environment from their release. Some active measures they use are constructing fences, limiting land use in the area particularly for farming, and constructing covers over nuclear waste that prevent seepage into groundwater, which also suppresses radiation and helps stabilize the waste, so it doesn't disperse through the soil. Mathematical modeling and numerical simulation of the repository and its radiation emissions are key here to ensure that the nuclear radiological risk to soil and groundwater remains low or contained in the immediate vicinity.

Moving forward, the future of chemistry in industry must be green that is, taking into account the lifecycle of hazardous chemicals and their ultimate fate and vector through the environment. There are many factors that go into this, but the primary objective is getting ahead of the curve when it comes to environmental protection. This will require using more benign replacements for certain harmful chemicals, reaction pathways that produce less wasteful and potentially harmful side-products, choosing more inherently safer designs in chemical processes to prevent accidents and spills, and producing more accurate mathematical modeling and sampling to better understand the final state of the chemical in the environment and the major vectors and sinks it will effect. At the same time, regulatory officials and industry must work together to refine methods and technologies for analyte monitoring and to limit pollution into the environment. Forward thinking and collaboration are not only required of industrialists and regulatory officials in the US, but also regulatory bodies the world over to combat the environmental challenges affecting all humans. A major step in this regard was the Paris Climate Accord (2016) that brought almost all nations together with the aim to reduce carbon emissions through individualized voluntary programs. This voluntary or peer-pressure approach is a critical innovation in getting all nations to participate, as previous mandatory emission standards proved to be roundly ineffective.

Key analytical techniques

Analytical chemistry can be applied in measuring the many pollutants that often end up in soil, water supplies, and the air we breathe. The essential in situ measurements commonly used across various methods and media include pH, temperature, turbidity, dissolved oxygen, and conductivity if ions are present. Table 2.4 provide a summary of various techniques organized into the three media discussed.

Analytical technique	Media of use	Analyte	Measurement	Drawbacks
Gravimetric analysis via precipitation	Ground, water	Metal ions, other various ions	Mass and concentration	Large amount of chemicals needed, often forms sludge-disposal, requires alkaline pH
Gas chromatography (GC) and high-pressure liquid chromatography (HPLC)	Ground, water, air	Gases (NO_x, SO_2, O_3, PMs) dissolved or in air, organics (PAH, VOCs, DMU from pesticides)	Presence/concentration	Very sensitive detection of multiple signals if highly contaminated samples
Thermogravimetric/Combustion Analysis (TGA)	Ground, air	Fossil fuels such as oil, gases, hydro- carbons, certain metal hydrates	Mass, amount	Can be highly flammable, not the most consistent with metal complexes (EDTA)
GC/MS	Ground, water, air	Virtually any compound which can be ionized in gaseous state (dioxins, PCBs, etc.)	Identity, presence, mass	Requires com- pound to be on database, must be volatile enough
Continued on next page				

Analytical technique	Media of use	Analyte	Measurement	Drawbacks
GC/MS	Ground, water, air	Virtually any compound which can be ionized in gaseous state (dioxins, PCBs, etc.)	Identity, presence, mass	Requires compound to be on database, must be volatile enough
Ion exchange	Ground, water	Most metal ions (Pb^{2+}, Cd^{2+}, Zn^{2+}, etc.)	Concentration, presence	Not best if large metal concentration, sensitive to pH, wastewater impurities
Electrolytic recovery	Water	Most metal ions	Mass, presence	Corrosion of electrodes, high current impurities
Absorption spectroscopy	Water, air	Most metal ions dissolved, dissolved gases (UV usually used)	Concentration, presence	Can be hard to interpret with many impurities, must use solvent blank
Extraction methods (DMU, w/methanol, Soxhlet, organic)	Ground, water	Mostly organic compound focus (PAH from soil), pesticides, aqueous	Presence, identity using GC/MS, NMR, etc. after extraction	Other contaminants may also be extracted
Adsorption techniques (zeolites, nanoparticles)	Ground, water	Mainly small metal atoms/ions, other nanoparticles	Identity, presence, concentration Impurities can affect adsorption	
Passive diffusion/active pumping	Water, air	Dissolved molecules, gases (NO_x, PAHs, etc.)	Concentration, presence	Impurities affect rates/concentrations of diffusion

Table 2.4: Major analytical techniques, media, and various drawbacks.

Notably, most of the analytes and pollutants depicted have various extraction methods that may then be followed by measurement techniques. Some of the analytical techniques depict both in tandem. In terms of extracting soluble metal ions from ground and water resources, there are both precipitant techniques and electroanalytical techniques. For example, often times basic titration using hydroxide ions with various samples of sewage water or water known to contain metal ions can help determine the concentration of the ions by means of gravimetric analysis of the precipitate. However, for convenience, such measures are often not taken, and an excess of base is added to simply precipitate out as many metal ions as possible. In other words, there may be a greater incentive to just clean the solution rather than taking analytical accounts of measuring concentration.

Another notable method of removing ions in contaminated wastewater is through ion exchange. However, this is usually limited to lesser metal ion concentrations. Matrices containing synthetic organic ion exchange resins are often used to exchange either cations or anions in solution. As stated though, high concentrations of metal ions can lead to contamination of the matrices by flooding them with organic wastes and other solids. For even purer solutions, such as processed water streams, electrolytic recovery (electrogravimetric analysis) can be used. This is a type of electroanalytical technique, which uses a current to deposit positively charged metallic ions onto the negatively charged cathode. The deposit can be measured to gravimetrically determine the relative mass and concentration of the metal. Nanoparticles are yet another method of extracting ions from water and soil, providing adsorbent surfaces capable of also being redox active. Further analysis can be conducted using GC/MS or a comparison of absorption spectra of the sample with various reagent/solvent blanks. In terms of organic wastes (VOCs, PCBs (Figure 2.3), dioxins (Figure 2.5), antibiotics, pesticides, etc.) mixed with aqueous water solutions, various methods of extraction with organic solvents can separate these from the metal ions. For example, 1,3-dimethylurea (DMU, Figure 2.7), a pesticide, can be extracted from the soil using methanol. Further cleaning of the solution followed by GC can determine the concentration and presence of DMU. In addition, combustion analysis of organic aliphatic hydrocarbons can help to determine the identity of unknown oils, natural gas pollutants, and fossil fuels.

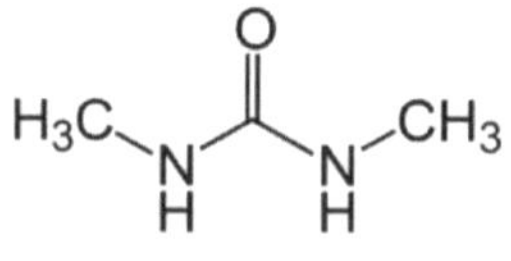

Figure 2.7: The structure of 1,3-dimethylurea (DMU).

Analysis of air is mainly completed by capturing the gas of interest either passively through diffusion or actively through a pump. Passive diffusion tends to be more cost effective, with a steel mesh filter covering the tube containing triethanolamine (TEA, Figure 2.8). This compound reacts with nitrogen oxides to form nitrites and is sent to the laboratory. Washing of the metal disc with water collects the nitrite ions in solution which can be examined with UV absorption: the amount of light absorbed can determine the amount and concentration of nitrogen oxides that were present in the air collected. Different diffusion tubes can be used for other organic compounds such as benzenes and PAHs. GC or HPLC can be used to further identify other various gases. Active pumping of gases into sealed containers are also used. These tend to measure various concentrations of molecules in the air as they change hourly in time. GC/MS can directly determine the amount and presence of sulfur trioxide, carbon monoxide, ozone, PMs, and other gases collected.

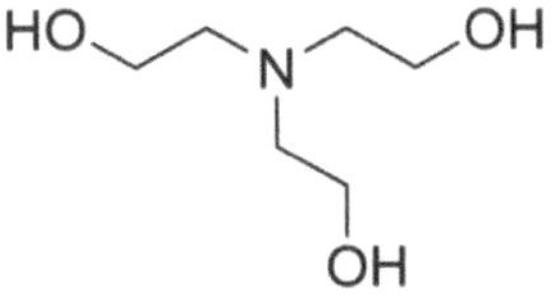

Figure 2.8: Structure of triethanolamine (TEA).

Conclusions and future work

When controlling pollution in industry, there are three main steps to be taken:

- The pollutants must be identified. For this purpose, identification processes must first be conducted on a small scale in a controlled environment before they can be scaled up and implemented on an industrial scale. Rigorous analysis of the raw materials as they are used or produced in various industries can lead to a better understanding of the pollutants involved and the extent of the environmental impact they may elicit. The detection of both predicted chemicals and unexpected ones spurs the development of new ideas for regulations. At the same time, successful analysis depends on the correct use of the many different analytical methods available for the detection of these environmental pollutants, whether they are found in the air, water, or ground.
- Once the pollutants among the products have been identified, it must be determined which analytical techniques should be used to detect it

in the environment. The nature of the pollutant is the principle factor in this decision, including how it enters the environment, where it is likely to be found, its physical state and chemical properties, and its effects on humans and the ecosystem at large. The nature of the particular environment should also be considered here, including geography and topography, rock and soil conditions, weather patterns, the presence or absence of water, and the lifeforms present, as these factors can affect both how pollutants impact the environment and how the pollutant can be detected.

- After the pollutants have been identified and the appropriate analytical methods have been determined, formal regulations may then be imposed. It is important to know what pollutants are produced by a particular industrial process or human activity before establishing regulations in order to make sure that nothing important is being missed and that no unnecessary restrictions are being imposed to control pollutants that are not actually being produced. It is also important that the analytical methods are both accurate and consistent so that industries are not regulated too tightly or too loosely.

It is also important to note that the analytical techniques presented herein are not entirely optimal or ideal. Many of the analytical methods described above involve chemical reactions, which require the consumption of reagents and other materials, yielding waste products as a result. On top of that, these techniques are easily hindered by impurities in the samples, so it would be in our best interest to find innovative, efficient ways to isolate substances of interest from impurities when conducting the analysis. For example, this line of innovation would improve spectroscopic methods by cleaning up the signals. Alternatively, one can study the substances denoted as impurities themselves in the samples and establish standardized algorithms, formulas, or methods for screening them out in the data.

In conclusion, the regulation of pollution in the environment is a continually evolving process. Many of the regulations in effect today developed over a relatively short period of time as people became increasingly aware of the impact of such pollution on public health. As new industries continue to develop, existing industries continue to adopt new practices and processes, and the world population expands, the types, nature, and impact of environmental pollution will continue to change over time. Knowing this, those responsible for regulating pollution must continue to develop and optimize analytical techniques in order to provide the necessary data to implement regulations in a manner agreeable to all involved.

Bibliography

Australian Department of the Environment and Energy. *Sulfur Dioxide Factsheet* (2005) http://www.environment.gov.au/protection/publications/factsheet-sulfur-dioxide-so2.

H. I. Abdel-Shafy and M. S. M. Mansour, A review on polycyclic aromatic hydrocarbons: Source, environmental impact, effect on human health and remediation. *Egypt. J. Pet.*, 2016, **25**, 107.

S. Chowdhury, M. A. J. Mazumder, O. Al-Attas, and T. Husain, *Sci. Total Environ.*, 2016, **569**, 570, 476.

EPA. *Particulate Matter (PM)* https://www.epa.gov/pm-pollution/particulate-matter-pm-basics#PM. .

EPA. *Summary of the Clean Air Act.* https://www.epa.gov/laws-regulations/summary-clean-air-act.

EPA. *Controlling Hazardous Air Pollutants*. https://www.epa.gov/haps/controlling-hazardous-air-pollutants.

EPA. *Criteria Air Pollutants: NAAQS Table.* https://www.epa.gov/criteria-air-pollutants/naaqs-table.

EPA. *Safe Drinking Water Act.* https://www.epa.gov/sdwa.

EPA. *Clean Water Act Compliance Monitoring.* https://www.epa.gov/compliance/clean-water-act-cwa-compliance-monitoring.

EPA. *Underground Injection Control.* https://www.epa.gov/uic/underground-injection-control-regulations-and-safe-drinking-water-act-provisions.

EPA. *Frank Lautenberg Chemical Safety for the 21st Century Act.* https://www.epa.gov/assessing-and-managing-chemicals-under-tsca/frank-r-lautenberg-chemical-safety-21st-century-act.

EPA. *Nuclear Waste Policy Act.* https://www.epa.gov/laws-regulations/summary-nuclear-waste-policy-act.

- EPA. *Radioactive Waste Disposal: An Environmental Perspective.* https://www.epa.gov/sites/production/files/2015-03/documents/000003ob.pdf.

S. M. Horvath, Nitrogen dioxide, pulmonary function, and respiratory disease. *Bull. N.Y. Acad. Med.*, 1980, **56**, 835.

K.-H. Kim, S. A. Jahan, E. Kabir, and R. J. C. Brown, A review of airborne polycyclic aromatic hydrocarbons (PAHs) and their human health effects. *Environ. Int.*, 2013, **60**, 71.

E. V. Lau, S. Gan, and H. K. Ng, Extraction techniques for polycyclic aromatic hydrocarbons in soils. *Int. J. Anal. Chem.*, 2010, **2010**, 398381.

J. L. Martinez, Environmental pollution by antibiotics and by antibiotic resistance determinants. *Environ. Pollut.*, 2009, **157**, 2893.

J. Müller and E. Rohbock, Method for measurement of polycyclic aromatic hydrocarbons in particulate matter in ambient air. *Talanta*, 1980, **27**, 673.

National Research Council, *Challenges for the Chemical Science in the 21ˢᵗ Century*, The National Academic Press, Washington, D.C., 2003.

R. H. Nuttall and D. M. Stalker, Limitations of thermogravimetric analysis of EDTA metal complexes as a method for structure determination. *J. Inorg. Nuclear Chem.*, 1978, **40**, 39.

M. B. Pell and J. Schneyer, *The thousands of U.S. locales where lead poisoning is worse than in Flint*. Reuters. December 19, 2016. https://www.reuters.com/investigates/special-report/usa-lead-testing.

C. Potera, Chemical exposures: cats as sentinel species. *Environ. Health Perspect.*, 2007, **115**, 580.

M. I. Sierra, A. Valdés, A. F. Fernández, R. Torrecillas, and M. F. Fraga, The effect of exposure to nanoparticles and nanomaterials on the mammalian epigenome. *Int. J. Nanomedicine*, 2016, **11**, 6297.

L. Zhang and M. Fang, Nanomaterials in pollution trace detection and environmental improvement. *Nanotoday*, 2010, **5**, 128.

Chapter 3: Chemical Industry

Sydney Stocks, Pavan M. V. Raja and Andrew R. Barron

Introduction

The chemical industry produces the most highly used chemicals worldwide at a large scale. These chemicals are in turn used for a wide variety of purposes. Many of the top produced chemicals are key ingredients in the manufacture of smaller-scale and more complex chemicals, and others go on to play roles in industries such as agriculture, food, and plastics. The large-scale of production in the chemical industry poses unique challenges, as do constraints imposed by government regulations affecting the industry. Purity, precision, energy consumption, and waste production must all be considered, as well as the analytical techniques used to determine these measures. This paper will explore the current chemical industry environment and go further to suggest future innovation.

Table 3.1 shows that sulfuric acid is the most produced chemical in the world, followed by nitrogen, ethylene and so forth. Table 3.1 also details the annual global production values of the top ten chemicals, and the most frequent and relevant applications for the use of each of the top ten worldwide chemicals. These uses range from making fertilizers to plastics to producing other important chemicals in the industry.

Chemical	Annual global production value	Applications
Sulfuric acid (H_2SO_4)	231 million tons (2011-2012)	Phosphate fertilizers, metal processing, phosphates, fibers, hydrofluoric acid (HF), paints, pigments, pulp, paper
Nitrogen	UK: about 5 million tons	Make ammonia (NH_3), create inert environment for food pack- aging, production of semiconductors, refrigeration
Ethylene (C_2H_4)	134 million tons (2014-2015)	Produce polyethylene (Figure 3.1a), PVC (Figure 3.1b), and polystyrene (Figure 3.1c), ethylene oxide (C_2H_4O), and ethanol (C_2H_5OH)
Oxygen	UK: about 5 million tons	Production of steel, copper, lead, clean sewage, produce chemicals, e.g., nitric acid (HNO_3)
Continued on next page		

Chemical	Annual global production value	Applications
Propylene (C_3H_6)	94 million tons	Produce polypropylene (Figure 3.1d), propenal (Figure 3.1e), acrylonitrile (Figure 3.1f), cumene ($C_6H_5{}^iPr$), propylene oxide (Figure 3.1g), butyraldehyde (Figure 3.1h)
Chlorine	65 million tons (2016)	Produce PVC and solvents, water purification, paper and pulp
1,2-dichloro-ethane (EDC, Figure 3.1i)	34.5 million tons (2013)	Produce PVC, produce solvents used in textile industry and metal cleaning
Phosphoric acid (H_3PO_4)	43 million tons (2014, 2016)	Produce fertilizers including monoammonium dihydrogenphosphate (MAP, $[NH_4][H_2PO_4]$), triple superphosphate (TSP, $[Ca(H_2PO_4)_2.H_2O]$), diammonium phoshate (DAP, $(NH_4)_2HPO_4$), soft drinks
Ammonia (NH_3)	146 million tons	Produce fertilizers including urea and ammonium salts, polyamides, nitric acid
Sodium hydroxide (NaOH)	70 million tons	Produce bleach, paper, soap, organic chemicals, use as cleaning agent

Table 3.1: Key materials used in the chemical industry.

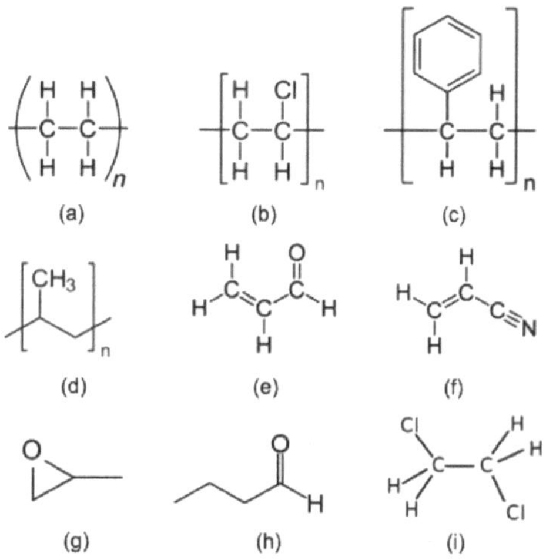

Figure 3.1: Structure of selected industrial scale chemicals: (a) polyethylene, (b) PVC, (c) polystyrene, (d) polypropylene, (e) propenal, (f) acrylonitrile, (g) cumene, (h) propylene oxide, and (i) ethylene dichloride.

Key analytical techniques

There are various techniques used to analyze different chemicals, each with their flown unique function. Table 3.2 briefly describes twelve common analytical techniques used in industry.

Technique	Function of technique
Elemental analysis	Determine quantity of element of interest within a molecule/material
Melting point analysis (mpt)	Qualitatively identify samples with at least 90% purity
Cyclic voltammetry (CV)	Any process that includes electron transfer
Thermogravimetric analysis (TGA)	Monitor thermal stability of compound
Gas chromatography (GC)	Separates stationary and mobile gas phases
High-performance liquid chromatography (HPLC)	Separate and identify components in a mixture
Ion chromatography	Separates ions based on retention rates
Infrared spectroscopy (FTIR)	Distinguish between compounds or identify specific compound by analyzing spectra
Raman spectroscopy (Raman)	Determines chemical species by detecting light interactions
UV-visible spectroscopy (UV-visible)	Obtain absorbance spectrum of compound
X-Ray diffraction (XRD)	Determine molecular and atomic crystal structure by diffracting X-ray beams
Transmission electron microscopy (TEM)	Form image through electron transmission

Table 3.2: List of common key techniques for analyzing chemicals and their functions.

Regulations

In the wake of concerns over environmental pollution, sentiments magnified after the publication of the popular novel Silent Spring, the EPA (Environmental Protection Agency) was founded in 1970 by the Nixon administration. These sentiments culminated in the celebration of the first earth day by millions of Americans later that year. The EPA has grown substantially in the years since then, especially due to role expansions and pollution goals stated in subsequent governmental legislations such as the Toxic Substances Control

Act, Pollution Prevention Act and Chemical Facility Anti-Terrorism Standards. Other notable acts, and perhaps more so than the aforementioned, are the Clean Air and Clean Water acts. The focus of this section will align mostly with the Toxic Substances Control Act (TSCA), Chemical Facility Anti-Terrorism Standards program (CFATS), and Pollution Prevention Act (PPA) though as these regulations have broader goals and impacts not specifically geared towards chemical production regulation.

TSCA

The Toxic Substances Control Act (TSCA) was passed in 1976 and gave the government the authority to mandate restrictions, testing requirements, reporting and record keeping over the chemicals industry. It should be noted, however, that food, drugs, cosmetics, and pesticides are outside of this regulation and are covered by other entities like the FDA (Food and Drug Administration). The important TSCA mandates include testing of any new chemical before sale, import, or new use to ascertain any foreseeable dangers to the chemical's introduction. Because of the TSCA, records of approximately 83,000 chemicals (and growing) are also documented in the TSCA Inventory, which includes a detailed indication measure of specific rules per chemical.

One example of this act, and its earliest major success, was the implementation of the Lead-Based Paint Program (1971), which, outside of requiring notice to anyone purchasing a property with lead-based paint, also mandated testing and reporting of lead content in paint and an acknowledgment of receipt among transaction relationships.

An important arm of the TSCA is the Good Laboratory Practices Standards Compliance Monitoring Program (GLPS). The GLPS program is the testing body that is able to audit a lab to test for compliance with the regulations set by a governmental act (GLPS is also used to enforce the FIFRA or Federal Insecticide, Fungicide, and Rodenticide Act). This body a rms that testing by a chemical producer is consistent with the quality, validity, and integrity that the TSCA requires.

As for the effectiveness of the TSCA, a 2017 study connected the 1970 act to a 10% decrease in pollution in the first three years of the act in counties that previously exceeded the pollution thresholds. The interesting part of the study, however, was that this reduced pollution caused a significant change in the lives of those in the county. The study attributes 1% yearly income gain and (in present value terms) a lifetime income gain of approximately $4,300 per individual.

The TSCA was also updated in June of 2016 with the Frank R. Lautenberg Chemical Safety for the 21st Century Act (Figure 3.2). This change lent more power to the EPA including the ability to regulate more effectively the 62,000 chemicals that were already on the market when the act first passed in 1970. In fact, under the previous rendition of the TSCA, even asbestos ($Mg_3Si_2O_5(OH)_4$, Figure 3.3) still passed regulation, so an updated TSCA was in order. Today, the EPA can pass regulation on specific chemicals much more rapidly, so toxic substances like asbestos can be removed from consumer chemicals and products much more quickly.

Figure 3.2: Author of the Chemical Safety for the 21st Century Act Frank Raleigh Lautenberg (1924 - 2013) who served as a United States Senator from New Jersey.

Figure 3.3: SEM image of asbestos fibers.

CFATS

The Chemical Facility Anti-Terrorism Standards program caters more specifically towards Chemicals of Interest (COIs) and factories that produce them. COIs are, in general, chemicals that are particularly toxic, flammable or corrosive. For example, nitromethane (Figure 3.4) is a COI because it can be configured into an Improvised Explosive Device (IED). The program is more di cult to describe generally because it adapts oversight to prevent release, theft, and sabotage of these chemicals. One consistent aspect of it is the Top Screen, which measures certain thresholds for a facility, providing information towards how best to implement the CFATS program. This screen ranges from testing connection to public water supply to susceptibility to foreign sabotage.

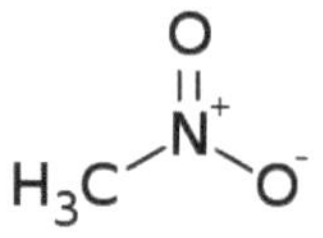

Figure 3.4: Structure of nitromethane.

PPA

The Pollution Prevention Act (1990) was intended at reducing pollution at the source rather than through treatment and disposal, like the act it iterated: The Resource Conservation and Recovery Act, that focused on waste disposal and recycling. Despite a vaguely worded law, with regard to specific chemical guidelines and requirements, and lofty pollution reduction goals, the EPA has outlined clear and measurable reductions across the 502 chemicals it monitors in every area of the United States (including territories). The law itself instead left much of the interpretation up to the regulatory bodies of the EPA that it created and has documented success in the available 2011 through 2015 data that has been made public.

Furthermore, more modern enactments like the PPA and Frank R Lautenberg Act (Figure 3.2) have lent the agency the power necessary to update standards and eliminate dangers like asbestos or volatile chemicals like nitromethane. Looking forward, however, against the current of international standards for pollution and emissions tightening, the United States has been producing policy to lift standards. The Trump administration has been vocal about disapproval of the previous administration's environmental policies, including the selection of Scott Pruitt as the new head of the EPA. Pruitt, who has previously spoken out against the existence of the EPA, is likely to slow down funding or progress towards tighter pollution standards in favor of more industrial freedom.

Challenges and need for innovation

In the chemical industry, a majority of the chemicals produced, those represented as top-selling chemicals, have long since been scaled up to mass production. As a result, many of the struggles the industry faces aren't the usual R&D roadblocks. Instead, the industry works hard to address the constant challenges of delivering consistent products at precise purities.

Purity is valued in almost any industry that uses mass-produced chemicals. Certain industries like electronics and pharmaceuticals place even further emphasis on these points, but almost every industry has written purity standards. Example standards are the Food Chemicals Codex, the United States Pharmacopeia National Formulary, and the American Chemical Society Standards. The chemical industry must work to deliver as many purity levels as possible. In some cases, they can; Thermo Fisher offers eighteen different solvent grades, ranging from basic laboratory use to Optima solvents with parts-per-trillion purity. Unfortunately, other substances aren't purified to so many levels, and, for example, Thermo Fisher only offers seven different purities of acids. Other chemicals cannot even reach desired purities due to issues such as complexing.

Just as important as purity is the precision of that figure. This presents two large challenges, avoiding manufacturing errors and perfecting analysis. Mentioned above, complexing can hurt the purity of a substance, and unintentional complexing with moisture in the atmosphere is one instance where error can be introduced into a batch. Errors can come from many places; any extraneous particles whether from the air, lab equipment, or final packaging can cause or prevent reactions, accidentally changing the final product. Diffusion or improper storage over time can also ruin the purity of reagents. Even less expected issues like excess factory vibrations have caused whole batches of polymer to form incorrectly. Human error can come in many ways, and it is another significant issue in processes that haven't yet been fully automated.

With abundant possible errors, analytical quality control is very important. When in a large-scale environment, this presents a specific challenge because although instruments are progressing towards using smaller amounts of sample, we need to generalize these results to one or many large batches. There are two different ways to approach these sampling issues: statistics or regulations. There is statistical theory that can be used on a general level to determine the best sampling procedures for any desired final purity. On an industry-by-industry level, there are regulations setting the sampling standards similar to those for purity standards. The happy medium probably takes both into account, and in the chemical manufacturing industry; different grades of chemical will follow different sampling techniques. The regulations will give an acceptable system for a basic, lab grade reagent, but statistics will probably be necessary to precisely create batches of increasingly higher purity.

Creating accurate and precise products is an everyday challenge for the industry, but looking forward, there is a need for a much more transformative improvement. Currently, the industry expends a lot of energy to deliver these consistent and accurate chemicals, and that is the largest way in which it could improve. Purity processes like distillation account for 10-15% of the world's energy consumption, and better purification techniques in oil and paper alone could save 100 million tons of carbon dioxide emissions and $4 billion in energy costs annually. This isn't just an opportunity for improvement; with new legislation like the Paris accord being put in place every day to tighten emissions regulations, major process improvement will likely soon be a necessity to keep the same manufacturing standards.

Conclusions and future directions

The chemicals industry is characterized by the mass production of some of the most important chemicals in the world. Its chemicals go on to play valuable parts in many further industries. The sheer volume of production necessary in the chemical industry shows just how important their role is in not just science, but in a vast array of fields. It is di cult to find an industry that is not somehow connected to the chemical industry. The most relevant regulation to the field is the TSCA, and additional legislation also aids in ensuring that the chemical industry is not only efficient but also safe. In addition to adjusting to meet regulations, improved purity and precision continue to be strived for in the industry. Key analytical techniques are used to measure these qualities in products and continue to be improved upon and specialized to certain fields. The chemical industry continues to grow, and it will continue to have a profound impact across the globe in years to come.

In the coming years, technological improvements will certainly be focused on improving purity for use in both pharmaceuticals and high-grade electronics, but this will come at a balance between the environmental cost of purification as well as the high cost at current standards. To address this, the industry is continually searching for new, creative methods of purification that will allow for less cost, waste or both. Along these lines, despite the effectiveness of procedures like HPLC, the high cost (including time investment and monetary) of developing standards for new applications of HPLC leave the industry researching again for new methods. On the topic of purification costs to the environment, perhaps the current US government will loosen environmental standards and allow for higher purification capacities, but international trends will likely outpace this domestic trend in the long run and regulate the industry in favor of cleaner natural resources.

Bibliography

A S Rathore, M. Yu, S. Yeboah, and A. Shama, Case study and application of process analytical technology (PAT) towards bioprocessing: Use of on-line high-performance liquid chromatography (HPLC) for making real-time pooling decisions for process chromatography. *Biotechnol. Bioeng.*, 2008, **100**, 306.

EPA: *Frank R Lautenberg Act*, https://www.epa.gov/assessing-and-managing-chemicals-under-tsca/frank-r-lautenberg-chemical-safety-21st-century-act.

EPA: *History of the Clean Water Act*, https://www.epa.gov/laws-regulations/history-clean-water-act.

EPA: *Summary of Clean Air Act*, https://www.epa.gov/laws-regulations/summary-clean-air-act.

D Harvey, *Modern Analytical Techniques*, McGraw Hill, New York (2000).

D C. Harris, *Quantitative Chemical Analysis*, 8th edn., W. H. Freeman and Company, New York (2010).

48

Chapter 4: Oil and Gas Industry

Kyle Chow, Thanh Huynh, Evan Rebesque, Reece Rosenthal,
Troy Tabarestani, Pavan M. V. Raja and Andrew R. Barron

Introduction

Fossil fuels come from dead plant and animal matter that were naturally buried under ancient waterways and land masses. Over the course of millions of years, the decaying biomass is subjected to extreme heat and pressure leaving behind trapped hydrocarbons (natural gas, crude oil, and kerogen). These hydrocarbons provide rich fuel sources that most modern technology depends upon. Due to the prehistoric pattern of ancient waterways and land masses, natural oil deposits were formed in various geological locations. This leads to some countries having disproportionately larger deposits of oil than others giving those countries major political and economic advantages. For example, Venezuela is estimated to have the largest deposit of oil with 300 billion barrels, followed by Saudi Arabia with 260 billion barrels while the US comes in 10th place with 40 billion barrels. On the other hand, Japan has only 40 million barrels and, therefore, is largely dependent on foreign sources.

Crude oil, or unrefined oil, is a mixture of volatile hydrocarbons that must be separated and treated to meet the quality and safety standards needed for consumers. When extracting and processing crude oil, oil companies are concerned with three stages of operation aptly named based on the distance from the source of oil:

- upstream (uses analytical techniques to find possible trapped oil deposits and analyzes sedimentary pro les to determine the most efficient extraction method),
- midstream (focuses on cost-effective ways of safely transporting crude oil that, often times, contains corrosive chemicals),
- downstream (deals with separation and purification of crude oil using techniques, such as fractional distillation and cracking).

The upstream sector of the oil and gas industry is primarily concerned with searching for crude oil or natural gas, and with bringing it to the surface once it is discovered. To find underground reservoirs that contain oil and gas, companies in the industry have developed advanced techniques to predict the locations of these reservoirs and visualize their sizes and geological properties. Using geophysical imaging technology, scientists elucidate where the reservoirs are and how valuable their contents may be. They then prepare an exploration plan to more precisely identify the size and contents of the reservoir. Exploration of offshore drilling sites in the USA is regulated by the Bureau of Ocean Energy Management. This agency was created by the US Department of the Interior to put environmental safeguards in place, while overseeing the management of offshore resources. Once a reservoir is identified and a well is drilled, oil and gas companies go to great lengths to optimize the recovery factor, or fraction of oil extracted, from the wells, thus maximizing profit. Some unconventional methods of extracting oil and gas include fracking, or high-pressure hydraulic fracturing, which is used to free natural gas from the shale rock it is deposited within.

Once the crude oil and gas has been collected, it must be refined into usable molecules in the midstream sector. Large hydrocarbon chains are broken up into smaller, more usable molecules in a process called cracking. Thermal cracking uses high pressures and temperatures, while catalytic cracking uses catalysts, to cleave carbon-carbon bonds and generate lighter oil fractions for separation. Oil refining often generates large volumes of hazardous waste, including benzene (C_6H_6), heavy metals, hydrogen sulfide (H_2S), acid gases, mercury, and dioxin, which cause environmental concerns that have still not been completely alleviated by the industry.

The downstream sector consists of determining the value of refined oil and gas fractions and selling them as finished products to other industries. Perhaps the most important downstream product is gasoline, which is used as a fuel in most internal combustion engines. Because of its widespread use, the contents of gasoline are strictly regulated by the EPA to safeguard our air quality. Not all gasoline is created equal, and the performance of a specific fraction of gasoline is rated by its octane level. Natural gas is another fossil fuel consisting primarily of methane, which is used mostly for the generation of electricity, as well as for generating heat in homes. Natural gas can provide energy at a higher efficiency than most fuel sources. Chemical products derived from oil and gas, which are not fuels, are generally referred to as petrochemicals. These can come in the form of detergents, fertilizers, medicines, paints, plastics, and even explosives. The large variety of uses for these petrochemicals, as well as the widespread use of fossil fuels, makes the oil and gas industry an industry of great importance to our society.

Materials in the oil and gas industry

Crude oil and petroleum found underground will always consist of a mixture of different hydrocarbons that vary in size (i.e., M_w) and arrangement (i.e., linear, branched, aromatic). Furthermore, the specific mixture of hydrocarbons obtained (and thus the crude oil's density) is also largely dependent on the source the oil was obtained from. Crude oil is classified by the American Petroleum Institute (API) and the SARA (saturate, aromatic, resin and asphaltene) analysis, which are based on physical and chemical characteristics, respectively. Table 4.9 provides a summary of both methods. Depending on impurities, crude oil can additionally be classified as sweet crude (<0.42% sulfur) or sour crude (>0.50% sulfur, often with other impurities like carbon dioxide).

Classification type	Method	Types
API gravity (physical)	Measure density of oil in comparison to water (higher density corresponds to lower API gravity)	Light crude: API 31.1 Medium crude: API 22.3 - 31.1 Heavy crude: API <22.3 Extra heavy crude: API <10
SARA (chemical)	Measure polarizability and polarity of oil	S (saturate): nonpolar, contains linear, branched, and cyclic saturated hydrocarbons A (aromatic): polarizable, contain aromatic rings R (resin): have polar substituents, miscible with heptane and pentane A (asphaltenes): have po-lar substituents, insoluble in heptane and pentane

Table 4.1: Classification of crude oil.

When processing crude oil, a wide variety of materials and chemicals are used in various stages of the process, a few of which are listed in Table 4.2.

Analytical chemistry is vital in each step of the production process, because being familiar with appropriate materials (examples in Table 4.3) is important in effectively and safely producing oil. This article attempts to present a bird's eye view of analytical chemistry in the context of the oil and gas sector, with an emphasis on the upstream aspect of this sector.

Material	Description	Common use
Metals	Metals are used in all parts of the oil/gas industry, from pipelines and drilling bits to storage tanks and processing equipment. Metals should be chosen such that they have the appropriate resistivity to stress, corrosion, or heat.	Steel, often carbon steel, is used that has good corrosion resistance and is thus used in most parts of production and processing. Steel alloys sometimes contain titanium or chromium to increase strength. Nickel alloys used in valves and piping above wellheads (Christmas tree). Copper is often used in valves and heat transfer applications due to cold-resistant properties.
Oil and water-based fluids/muds	When drilling into the Earth, fluids called muds are used to control pressure, which stabilizes rock and facilitates carrying of cuttings to surface.	Water-based muds (WBM) use water as the continuous phase, allowing for a cheaper alternative, but it can cause instability in shales. Oil-based muds (OBM) provide good stability and lubrication, but more expensive and potentially environmentally harmful. Air can sometimes be useful in drilling boreholes but is weak in supporting bore- holes and preventing fluids from entering.
Proppants	Proppants are sand-like materials suspended in water or other fluids.	Proppants are very useful in hydraulic fracking to keep fissures open. Choice of size, geometry, and weight can lead to improved conductivity, crush resistance, and acid solubility.
Fluid catalytic cracking material	In the fluid catalytic cracking (FCC) process, certain materials can be used to assist conversion of crude oil to refined products.	When larger hydrocarbons are broken down into smaller ones, the catalytic fluid reaction be- comes more important. It is critical that the catalyst be able to withstand impact, temperature, pressure, and actions of potentially poisonous metals.
Liquefied natural gas (LPG)	LPG can be produced by cooling natural gas to -162 °C, at atmospheric pressure, decreasing its volume by a factor of around 600.	LPG is very useful in transportation of natural gas across long distances or across deep water from remote locations. How- ever, LPG operations are expensive to run due to the maintenance of cryogenic tanks.

Table 11.10: Materials used in the oil industry.

Materials and chemicals	Annual global production	Major companies and groups	Purpose/application in the oil industry
HCl	2,800 tons	Dow Chemical, BASF, Bayer Corp, Oxychem	Used in hydraulic fracking as a corrosive agent to dissolve porous sedimentary rocks
Modified sands	45 million tons (2017)	CARBO Ceramics, High Crush LP, Emerge Energy Services, US Silica	Used in hydraulic fracking as a proppant to hold cracks open, allowing drainage of trapped natural gas and oil
Asphaltenes	82.76 million barrels/day[a]	United States, Russia, OPEC	Abundant crude oil component used to determine what analytical methods are needed
Paraffin wax	82.76 million barrels/day[a]	United States, Russia, OPEC	Dense portion of crude oil that must be precipitated out before further analysis
Naphthenes	82.76 million barrels/day[a]	United States, Russia, OPEC	Cyclic para ns that impact the density and type of crude oil
Polyaromatic hydrocarbons	82.76 million barrels/day[a]	United States, Russia, OPEC	Crude oil components that can serve as biomarkers
H_2S	9.6 million tons (2000)	Bechtel, Intertek	Found in crude oil and natural gas deposits. A common, unwanted corrosive chemical at high concentration.
Glass beads	-	-	Improves separation in fractional distillation technique by enhancing evaporation and condensation

Table 4.3: The use and production of materials encountered in the oil industry. [a]Quantity refers to total amount of crude oil produced globally.

Needs and challenges of the industry

In the US the oil industry is dominated by the multinational oil giants such as Exxon, Shell, and BP. Internationally, OPEC, an organization of 15 major oil-exporting countries, heavily influences the world's market. All of these companies and countries are driven by profit. Among other factors, such as consumer habits and governmental regulations, pro t margin is largely increased by lowering cost of production.

The cost of production is dependent on many factors including methods of extraction/production, regulations and taxes, quantity of available oil to be extracted, etc. For example, Saudi Arabia is able to produce a barrel of oil for only $10 per barrel due to the country's choice of regulations, taxes, and, of course, rich oil reserves. On the other hand, the US's cost of production is about $23 per barrel, but recent estimates put it at above $60. High production cost becomes a major problem when the market price of oil drops. For instance, the market price of a barrel of oil in the US plummeted to $25 toward the end of 2015 causing a loss of hundreds of thousands of industry-related jobs and cuts in R&D funding.

The main challenge facing the oil and gas industry is not a scientific one; instead, it is the political (and intertwined economic) aspect of the oil industry that is most complicated. Oil, in some cases, is power. The price of oil in the consumer market fluctuates constantly due to the actions of OPEC members and other oil giants around the world fluctuations that can be a response to political events, new discoveries of domestic oil, or otherwise non-scientific reasons. Therefore, it is the job of the scientist to ensure that oil extraction and purification continues in an efficient, environmentally friendly manner, even though the political instability of the present.

In order to continue drilling in the United States, the oil industry must adhere to several environmental regulations that have the overall effect of raising the cost of oil production. These laws and restrictions are divided into three main groups: laws regarding pollution, laws regarding waste, and laws regarding overall safety regulations related to the oil and gas industry. An overview of some of the critical federal laws governing the oil and gas industry in the United States can be seen in Table 4.4.

One of the greatest challenges of the oil and gas industry is environmental impact. It is not an unknown fact that oil and gas byproducts have tremendous effects on the environment, and as a result more and more regulations on environmental impact are being put out. An example is the Clean Air Act (Table 4.5). Other than the Clean Air Act, there has been a move towards carbon tax, which is when industries must pay a fee on the burning of carbon-based fuels. Its job is to act as a monetary disincentive that can motivate a switch to clean energy. Another form of regulation of environmental impact is carbon sequestration, which involves finding some way to capture CO_2 gas as a solid or liquid form to be trapped in oceans or in geological structures, preventing it from being emitted. Such plans will require heavy planning and risk management, as well as analytical methods to keep track how much CO_2 there is and monitor potential leakages or distributions. Oil companies are actively trying to curb their emissions and reduce their footprint in order to meet various requirements, but it is also necessary to have accurate real time monitoring on the release of CO_2 that can keep the industry accountable.

Law	Regulation
Oil Protection Act (OPA)	Mandates that companies meet relatively strict standards when it comes to the prevention and response to oil spills and other forms of pollution.
Federal Clean Air Act (FCAA)	Regulates the amount of waste, and type of waste, that oil companies can release into the atmosphere as gas during the refining or extraction processes.
Clean Water Act (CWA)	Prohibits the contamination of drinking water tables or rivers, streams, ponds, or lakes by drilling processes.
Comprehensive Environmental Response, Compensation, and Liability Act (CERCLA)	Firmly places responsibility and legal liability for oil and gas spill cleanup in the hands of oil and gas companies.

Table 4.4: US federal oil and gas regulations.

Regulations	Regulation/guideline type
Natural gas transmission and storage facilities	National Emission Standards for Hazardous Air Pollutants (NESHAP)
Oil and natural gas production facilities	NESHAP
Crude oil and natural gas production, transmission and distribution	New Source Performance Standards (NSPS)
Crude oil and natural gas facilities	NSPS
Sulfur dioxide (SO_2) emissions from onshore natural gas processing	NSPS
Equipment leaks of volatile organic compounds (VOC) from onshore natural gas processing plants	NSPS
Control of volatile organic compound equipment leaks from natural gas/gasoline processing plants (1983)	Control Techniques Guidelines (CTG)

Table 4.5: Clean Air Act standards and guidelines for the oil and natural gas industry.

Key analytical techniques

Oil and gas mixtures are comprised of hundreds, if not thousands of different components, with each of these components reacting and behaving in a multitude of unique ways. In order to classify this abundance of compounds, an equally abundant number of analytical methods have been devised to help us understand the chemicals that we drill for every day. Many analytical methods focus on physical distinctions between molecules. The ability to identify molecules and their quantity is one of the most valuable techniques in all of analytical chemistry. Properties such as density, viscosity, and boiling point are some of the most accessible methods of classifying components in mixtures; however, in the majority of cases, complex oil and gas mixtures cannot easily be separated into their constituent components. As a result, more precise techniques must be used to help analyze oil. The most popular options are spectroscopy and chromatography, two techniques that are used in research all over the world.

One example of spectroscopy usage in oil and gas is to analyze asphaltene (Figure 4.1) content in crude oil. Asphaltenes are molecular substances consisting of hydrocarbons. If left undisturbed, asphaltenes form solids that have a tendency to clog pipes, damage equipment, and introduce themselves into processes that they do not belong in. In order to predict the precipitation of asphaltenes in crude oil, researchers have devised a method of determining the onset of solid formation. The indirect method is a technique that involves conducting UV-visible spectroscopy on a sample of crude oil. By comparing the absorbance of light and the age of crude oil, researchers are able to correlate the aggregation time of asphaltenes to the quantity of asphaltenes. The indirect method is one of the many techniques used to help researchers design and maintain oil and gas flows, and it would not be possible without the help of spectroscopy.

Chromatography is an integral part of the oil and gas industry as well. During a process known as visbreaking, long hydrocarbon chains are cracked into smaller chains, allowing for the production of more valuable types of oil (Figure 4.2). In order to classify the contents of visbroken oil, researchers use thin-layer chromatography (TLC). Thin-layer chromatography is performed on a thin plate, which is coated with a thin adsorbent (the stationary phase). Once a mixture is introduced to its surface, capillary action draws the liquid upwards, allowing for separation. By using TLC in conjunction with densitometry, one is able to identify the contents of visbroken oil.

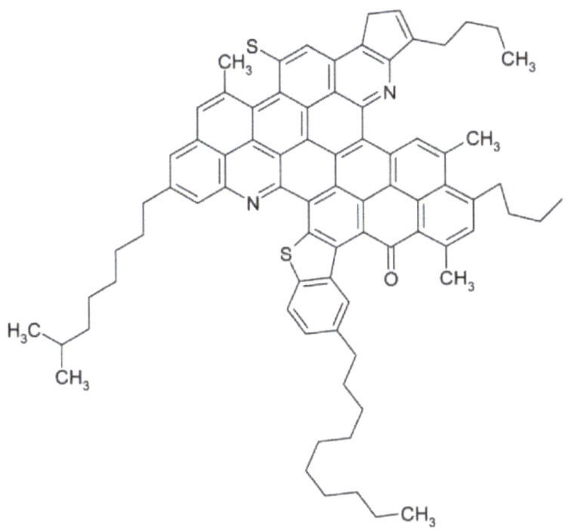

Figure 4.1: The idealized structure of asphaltene showing the types of functional groups that are present in a real sample.

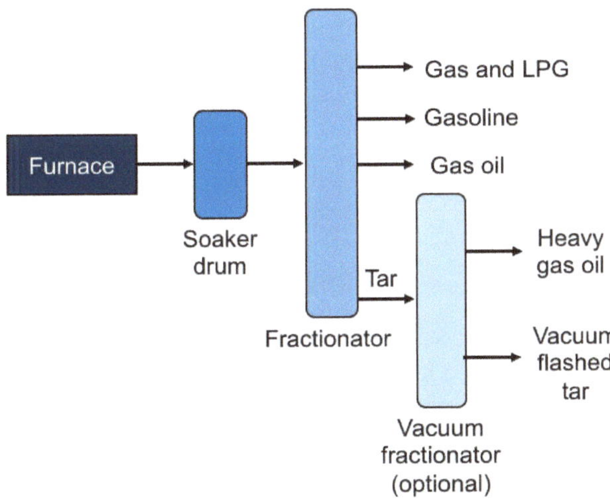

Figure 4.2: A schematic diagram of a visbreaker unit.

In the not-so-distant past, the methodology associated with reservoir exploration and subsequent crude oil extraction had been dominated by geochemists due to their advanced knowledge of topography and sediment analysis. However, over the past decades, the field of analytical chemistry has seen some revolutionary advancements that have brought new methods to the forefront of the industry. Table 4.6 briefly describes eight analytical techniques used for exploration/extraction, separation, and analysis of oil.

Method	Purpose/application of technique
Downhole fluid analysis (DFA)	Allows for in situ analysis of reservoir samples to determine composition of oil column
2D gas chromatography (2D GC)	Gives enhanced peak capacity, sensitivity, and resolution due to multiple separation dimensions
High-performance liquid chromatography (HPLC)	Assists in determination and separation of the main structural components of crude oil
Fourier-transform ion cyclotron resonance mass spectrometry (FT-ICR MS)	Detects the compositional complexity of the acidic/basic species like N-containing compounds at the molecular level
Inductively coupled plasma mass spectrometry (ICP-MS)	Determines the concentration of metallic elements and organometallic species in a sample
Ion mobility mass spectrometry (IM-MS)	Separates compounds based on their structural characteristics and charge state
Small angle X-ray scattering (SAXS)	Utilizes electron density differences and X-ray beams to collect structural information
Liquid crystal analysis	Useful for freezing point determination and ne-tuning partition methods

Table 4.6: List of common key techniques and their function for analyzing crude oil.

Extraction methods

The first and possibly most vital step for any oil company is to decide where they are going to drill and why. The importance of this decision on location will determine the productivity, pro t, and expenditure for the company. For this reason, much thought is placed on oil analysis before any drilling has even begun. Analytical chemists have brought a variety of techniques to this field; however, one of the simplest yet most effective is the downhole fluid analysis, or DFA, method. The process itself is very simple, a probe is sent down into the earth and is brought up with a vertical sample nearly 700 ft in length containing a variety of sediment and layers. With this column, a variety of information can be determined, especially with respect to biodegradation: by comparing various DFA samples from around the world, a company can predict the quality of the crude oil, compare compounds that are usually indicative of increased levels of biodegradation, and paint a picture of the oil's molecular components. By coupling this technique with other, more analysis-focused methods, an even greater pool of information can be obtained to gain a more holistic, predictive image of the crude oil that is going to be extracted from the potential reservoir.

Separation methods

Since crude oil is a complex mixture containing some compounds that cannot be analyzed by higher-end analytical methods such as mass spectrometry and liquid crystal analysis, separation must be used to eliminate certain molecules from crude oil before further analysis can be conducted. One manner of removing impurities is by precipitating out wax of lengthy carbon chains called para n wax that is too dense and not volatile enough for use in gas chromatography (GC). Another method of separation that has seen advances tailored toward the oil industry is multidimensional gas chromatography. When a single column does not offer the separation or resolution desired among a variety of crude-oil components, several columns, each with a unique stationary phase, are used in conjunction in order to isolate compounds from oil based off of different physical properties. Paired with flame ionization detection, multidimensional GC is so sensitive that miniscule amounts of matter can be detected in situations such as dilute-sample analysis of oil following oceanic oil spills.

If multidimensional GC is still insufficient in separating out compounds, high-performance liquid chromatography (HPLC) can serve as a precursor to GC. A novel method of HPLC uses a silver-modified stationary phase to separate saturated hydrocarbons from unsaturated ones. Another HPLC approach uses a hyper-crosslinked stationary phase formed from alkyl bonds between aromatic residues of stationary-phase molecules; this hyper-crosslinked structure has been found ideal for separation of polyaromatic hydrocarbons as well as for separation of different nitrogen-containing aromatic compounds such as pyrrole (Figure 4.3a) and pyridine (Figure 4.3b).

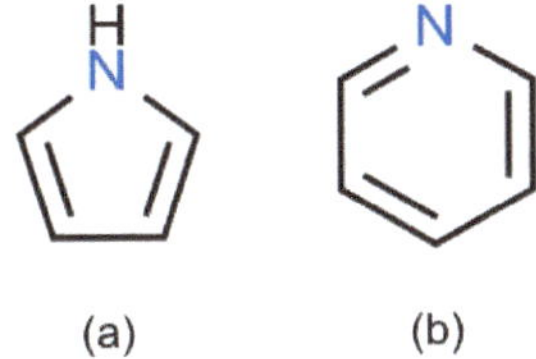

Figure 4.3: Structure of (a) pyrrole and (b) pyridine.

Molecular-level analysis methods

Subsequent analysis of the separated components is the next key step in the process. The main and most efficient method so far has proven to be FT-ICR coupled with MS. Using this method, incredible resolution of complex molecules can be attained due to a complex mechanism: by measuring the frequency of ion packets created by the machine, researchers are able to pinpoint a fairly precise range of weights for asphaltenes (350 - 1050 Da), something that previous methods of analysis would not have been able to determine with such high accuracy.

Even more sensitive is the ICP-MS method which uses extremely high temperatures to ionize almost everything in a sample. Using this technique, researchers are now able to accurately determine the concentrations of a variety of metallic elements and organometallics in their samples to ensure no water waste contains harmful pollutants. For more structural analysis, methods like ion mobility mass spectrometry (IM-MS) can be implemented which use a carbon nanopowder to separate compounds based not only on mass/charge ratios, but also structural features like having linear, branched, or condensed 3D shapes.

Finally, if an even more detailed characterization is needed, methods like small angle X-ray scattering (SAXS) and/or liquid crystal analysis can be used to discover crucial information about specific components of crude oil, specifically asphaltenes. With this additional data, future methods can be tailor-made for samples of crude oil from specific regions of the world that share common structural characteristics to increase cost-effectiveness and thereby profitability for the industry.

Conclusions and future directions

With an expanding market for oil and an accelerated need to quickly locate and characterize deposits of crude oil around the world, the work of analytical chemistry in the realm of oil and gas is becoming more paramount as time progresses and is constantly seeing novel opportunities for improvement. Throughout the last century, chemists working with oil have established a reliable system of first carrying out preliminary sample assessment through downhole fluid analysis, then separating large classes of crude oil components from each other using gas and high-performance liquid chromatography, and finally implementing specific analytical tools such as mass spectrometry and inductively coupled plasma in order to characterize oil based of structural details and varying levels of nitrogenous and organometallic molecules. In response to harsher regulations and safety standards placed on the processing and handling of petroleum, the oil industry will have to continue to enhance the detection range of its analytical instruments in coming years so that minute quantities of carcinogens, poisons, and other harmful chemicals can be identified in drilling wastes or before petroleum products are certified for quality and/or are approved for commercial use.

As demand for oil and gas continues to grow, the industry must rise to meet new challenges in production. New sources of fuel and new methods to find these sources have created an industry that is dynamic and ever-changing. Existing technology is effective but has proven to be inefficient and redundant with severe limitations in terms of adaptability and user-friendliness. Therefore, it is extremely important that new technology of the kind discussed above is developed in the short term in order to meet the projections of oil production, and in the long term, it is important that we shift toward improving the efficacy of extraction and utilization oil resources, given that they are not renewable. As the world becomes more environmentally conscious, it will be extremely important for the oil & gas industry to also become more mindful of the environment and analytical techniques such as those discussed in this article will go a long way in attaining related goals.

Bibliography

C. Becker, F. A. Fernandez-Lima, and D. H. Russell, Ion mobility-mass spectrometry: a tool for characterizing the petroleome. *Spectroscopy*, 2009, **24**, 38.

V. Cebolla, L. Membrado, M. Domingo, P. Henrion, R. Garriga, P. Gonzalez, F. Cossio, A. Arrieta, and J. Vela, Quantitative applications of fluorescence and ultraviolet scanning densitometry for compositional analysis of petroleum products in thin-layer chromatography. *J. Chromatogr. Sci.*, 1999, **37**, 219.

H. Devold, Oil and Gas Production Handbook: An Introduction to Oil and Gas Production, Lulu.com (2013).

EPA: *Oil and Gas Extraction Sector (NAICS 211)*, https://www.epa.gov/regulatory-information-sector/oil-and-gas-extraction-sector-naics-211,

F. A. Fernandez-Lima, C. Becker, A. M. McKenna, R. P. Rodgers, A. G. Marshall, and D. H. Russell, Petroleum crude oil characterization by IMS-MS and FTICR MS. *Anal. Chem.*, 2009, **81**, 9941.

M. C. Kennicutt II, The effect of biodegradation on crude oil bulk and molecular composition. *Oil Chem. Pollute.*, 1988, **4**, 89.

H. Longwell, The future of the oil and gas industry: past approaches, new challenges. *World Energy*, 2002, **5**, 100.

S. J. Maguire-Boyle and A. R. Barron, Organic compounds in produced waters from shale gas wells, *Environ. Sci.: Processes Impacts*, 2014, **16**, 2237.

S. J. Maguire-Boyle, D. J. Garner, J. E. Heimann, L. Gao, A. W. Orbaek, and A. R. Barron, Automated method for determining the flow of surface functionalized nanoparticles through a hydraulically fractured mineral formation using plasmonic silver nanoparticles. *Environ. Sci.: Processes Impacts*, 2014, **16**, 220.

D. Mao and V. Weghe, High-performance liquid chromatography fractionation using a silver-modified column followed by two-dimensional comprehensive gas chromatography for detailed group-type characterization of oils and oil pollutions. *J. Chromatogr., A*, 2008, **1179**, 33.

T. A. Maryutina and A. V. Soin, Novel approach to the elemental analysis of crude and diesel oil. *Anal. Chem.*, 2009, **81**, 5896.

P. Motamedi, H. Bargozin and P. Pourafshary, Management of implementation of nanotechnology in upstream oil industry: an analytic hierarchy process analysis. *J. Energy Resour. Technol.*, 2018, **140**, 052908.

O. C. Mullins, G. T Ventura, and R. K. Nelson, Visible–near-infrared spectroscopy by downhole fluid analysis coupled with comprehensive two-dimensional gas chromatography to address oil reservoir complexity. *Energy Fuels*, 2008, **22**, 496.

D. O'Rourke and S. Connolly, Just oil? The distribution of environmental and social impacts of oil production and consumption. *Environ. Resour.*, 2003, **28**, 587.

L. Petrakis, D. M. Jewell, W. F. Benusa, in *Petroleum in the Marine Environment*, Eds:L. Petrakis, F. T. Weiss, ACS Washington DC (1980).

B. Plainchont, P. Berruyer, J.-N. Dumez, S. Jannin, and P. Giraudeau, Dynamic nuclear polarization opens new perspectives for NMR spectroscopy in analytical chemistry. *Anal. Chem.*, 2018, **90**, 3639.

K. Qian, R. Rodgers, C. L. Hendrickson, and M. R. Emmett, Resolution and identification of elemental compositions for more than 3000 crude acids in heavy petroleum by negative-ion microelectrospray high-field fourier transform ion cyclotron resonance mass spectrometry. *Energy Fuels*, 2001, **15**, 492.

V. Rao and R. Knight, Sustainable Shale Oil and Gas: Analytical Chemistry, Geochemistry, and Biochemistry Methods, Elsevier, United Kingdom (2017).

R. Rodgers and A. McKenna, Petroleum analysis. *Anal. Chem.*, 2011, **83**, 4665.

E. Y. Sheu, Small angle scattering and asphaltenes. *J. Phys.: Condens. Matter.*, 2006, **18**, S2485.

M. Li, S. R. Larter, D. Stoddart, and M. Bjoroey, Liquid chromatographic separation schemes for pyrrole and pyridine nitrogen aromatic heterocycle fractions from crude oils suitable for rapid characterization of geochemical samples. *Anal. Chem.*, 1992, **64**, 1337.

R. Valdmanis, *As Trump targets energy rules, oil companies downplay their impact*, Reuters, https://www.reuters.com/article/us-usa-trump-oil-regulation-insight-idUSKBN16U1A9.

E. T. C. Vogt and B. M. Weckhuysen, Fluid catalytic cracking: recent developments on the grand old lady of zeolite catalysis. *Chem. Soc. Rev.*, 2015, **44**, 7342.

T. F. Yen, J. G. Erdman, and S. S. Pollack, Investigation of the structure of petroleum asphaltenes by X-ray diffraction. *Anal. Chem.*, 1961, **33**, 1587.

Chapter 5: Pharmaceutical Industry

Adrish Anand, Patrick Gilliam, Milad Najafabadi, Zoe Punske,
Pavan M. V. Raja and Andrew R. Barron

Introduction

The pharmaceutical industry discovers, develops, and sells a variety of drugs for use as medicine. These drugs are varied in nature and can be both self-administered or administered to a patient to alleviate countless ailments or diseases. Pharmaceutical corporations can produce both branded and generic products for the market. Pharmaceuticals have rapidly advanced in recent history, as much of the medicine available during humanity's existence comprised of folk remedies and products from apothecaries. In the past two centuries, medicines have quickly shifted from plant-derived botanicals to synthetically produced products.

The pharmaceutical industry plays a pivotal role in society and is responsible for saving countless lives, alleviating numerous symptoms, and contributing to the increased longevity of people. In short, many put their lives in the hands of the pharmaceutical industry. Given the critical role drugs play in modern society, they are subjected to myriad laws and stringent quality standards by federal governments and global organizations. To verify drug purity and efficacy, many analytical methods are employed. This article highlights the importance of analytical chemistry in supporting the pharmaceutical industry.

Key chemicals and materials

Table 5.1 shows the highest selling drugs in the pharmaceutical industry as of 2015. These drugs are used to treat a wide array of conditions including diabetes, cancer, and infectious diseases such as HIV. The active ingredients are organic compounds and biopharmaceuticals such as antibodies. With patents expiring and new drugs constantly in development, this list is in flux from year to year. As of Q3 2018, the largest pharmaceutical company by revenue is Johnson & Johnson followed by Roche and Pfizer.

Drug name	Active ingredient	Main indication	Company	2015 Revenue in $billions
Humira	Adalimumab	Immunology	AbbVie	14.0
Harvoni	Ledipasvir (Figure 5.1a) and sofosbuvir (Figure 5.1b)	Infectious diseases	Gilead Sciences	13.8
Enbrel	Etanercept	Immunology	Amgen/Pfizer	8.6
Remicade	Infliximab	Oncology	Johnson & Johnson/Merck	8.3
MadThera	Rituximab	Diabetes	Roche	7.1
Lantus	Insulin glargine	Oncology	Sano	7.0
Avastin	Bevacizumab	Oncology	Roche	6.7
Herceptin	Trastuzumab	Blood disorders	Roche	6.6
Revlimid	Lenalidomide (Figure 5.1c)	Infectious diseases	Celgene Corporation	5.8
Sovaldi	Sofosbuvir (Figure 5.1b)	Respiratory disorders	Gilead Sciences	5.2

Table 5.1: Highest-selling drugs as of 2015.

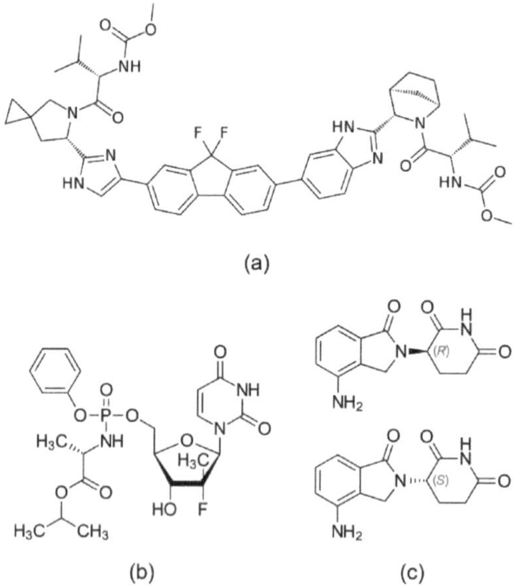

Figure 5.1: Structures of (a) ledipasvir, (b) sofosbuvir, and the two enantiomers of lenalidomide.

Table 5.2 gives the typical uses of selected chemicals and materials employed in the pharmaceutical industry.

Chemical/material	Application
Water	Medium
Silica	Applied to TLC plates, binder
Halogens	Stabilize bonding interactions
Vitamins	Biological basis for drug derivatives
Biological hormones	Biological basis for drug derivatives
Palladium	Mediates many drug syntheses
Wormwood	Natural plant containing medicinal extracts
Alumina	Common stationary phase
Ethyl acetate (Figure 5.2), hexane (C_6H_{14})	Common mobile phase

Table 5.2: Uses of selected chemicals and materials in the pharmaceutical industry.

Figure 5.2: Structure of ethyl acetate ($C_2H_5C(O)OCH_3$).

Challenges faced by the industry

The pharmaceutical industry faces several challenges in the market and in the laboratory. The biggest problem faced by the industry is reduced R&D budgets in recent times. An industry shift towards biological treatments such as proteins and antibodies and away from molecules has seen the cost and time associated with producing a single drug skyrocket. In a recent study 49.1% of the industry claimed that the drug discovery stage associated with high risk and high cost is the reason why they have cut spending on R&D. On average costing them over $2.5 billion per new drug, pharmaceutical companies are less incentivized to continue doing R&D themselves, and would rather just buy out smaller companies that do the R&D. In addition, return on R&D investments have declined from 10% to 3.2% in the last 8 years.

These factors directly impact the job of analytical chemists in the industry. With reduced budgets, cheaper analytical methods have to be used in order to separate and characterize compounds. It may be difficult for a company purchase expensive HPLC, LC-MS, and NMR instruments along with hiring the personnel to use them correctly and take care of them. Analytical chemists are still required to produce the same quality and reproducibility, but with cheaper/simpler technology. Consequently, this increases the time spent preparing samples and reduces the quality of the analysis. Analytical chemists face a large burden that is only growing as staff and equipment are reduced. A specific problem analytical chemists face with less funding is separating macromolecules that have very little heterogeneity. During a synthesis, the target molecule is produced along with several side products. The more complex a module becomes, the more complex the side products are. Additionally, side products look almost identical to the target drug (e.g., isomers) making them very di cult to separate from each other. Highly specific, expensive separating methods such as chiral columns or solvent and pressure gradient HPLC might be required to separate these compounds but with less funding it can be nearly impossible to separate such similar molecules.

Another problem drug companies face is the use of excipients in pills and tablets. Excipients are chemicals that are not the active ingredient and are used to enhance drug stability and function. In the past, they have been thought to be inactive; however, recently there have been several cases of adverse reactions from patients to excipients. This results in huge health risks and massive costs for the industry. For example, the compound tartrazine (Figure 5.3) is a coloring agent that gives pills a yellow color. This excipient is inactive in the majority of the population but can cause hyperactivity in some children. Other common excipients and their side effects are given in Table 5.3.

Excipient	**Function**	**Caution in Practice**
Tartrazine (Figure 5.3a)	Coloring agent	Hyperactivity in children
Aspartame (Figure 5.3b)	Sweetener	Caution in patients with phenylketouria
Benzalkonium chloride (Figure 5.3c)	Preservative	Bronchoconstriction, toxic for the eye
Sodium metabisulphite (Figure 5.3d)	Antioxidant	Bronchospasm, anaphylaxis
Propyl gallate (Figure 5.3e)	Antioxidant	Sensitive to skin
Lactose (Figure 5.3f)	Tablet filler	Caution in patients with galactosaemia
Sesame oil	Oil consistency	Hypersensitivity
Lanolin	Emulsifier	Sensitive to skin

Table 5.3: Common excipients and their adverse health outcomes.

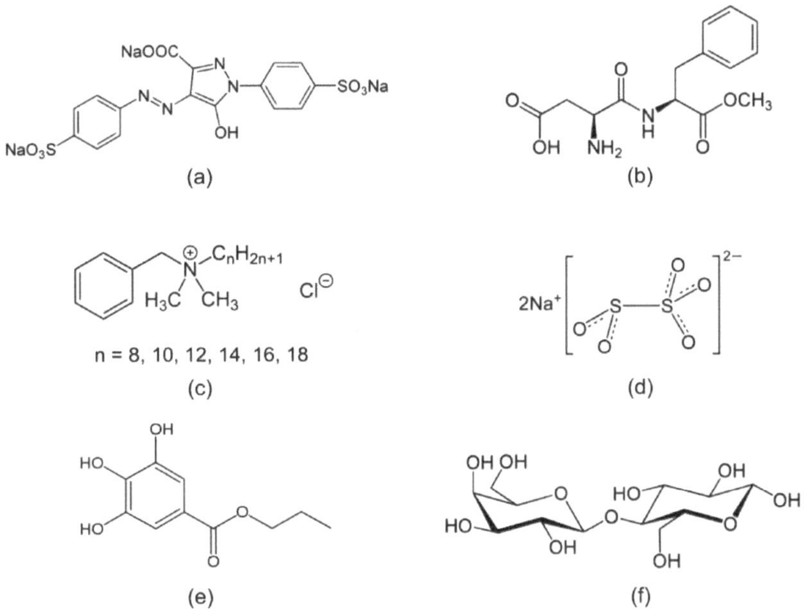

Figure 5.3: **Structures of common excipients: (a) tartrazine, (b) aspartame, (c) benzalkonium chloride, (d) sodium metabisulphite, (e) propyl gallate, and (f) lactose.**

Increasing innovation in R&D and drug delivery can reduce the time and cost associated with developing a drug. The industry needs to focus on using more technology and big data in order to make this move. With this focus the industry can see higher return on investments and increased budgets in R&D. Finally, overcoming these problems will improve the lives of analytical and synthetic chemists.

Regulations

Drugs are a special type of consumer product where consumers have very little knowledge of which drugs to take, how much to take, when to take them, and the adverse side effects of some drugs. Likewise, for the producers, immense knowledge about pharmacology and microbiology are necessary in ensuring the quality of a drug. Due to these strict requirements drugs are a product that face heavy regulations. The Food and Drug Administration (FDA) is the main regulator of drug companies in the US and enacts several measures to guarantee safety and quality for the consumer. The FDA enacts

Good Practice laws (GxP), which provide specifications for laboratory environments (GLP), clinical methods (GCP), manufacturing practices (GMP), and distribution of drugs (GDP). To protect the producer, the FDA enacts patents that can incentivize research and development (R&D) and provide massive profits for a company. Patent regulations vary depending on the category of drug such as new innovative, generic, or pediatric. Furthermore, to incentivize R&D into less common diseases, the federal government has passed the Orphan Drug Act (1983) and the Rare Disease Act (2002). The goal of these regulations is to ensure quality and safety for the consumer, but at the same time to allow for pro table R&D for the producer.

US Food and Drug Administration (FDA)

In the US the FDA regulates most food, human and animal drugs, therapeutic agents of biological origin, medical devices, radiation-emitting products for consumer, medical, occupational use, cosmetics, and animal feed. The history of the FDA can be traced to the latter part of the 19th Century when research began into the adulteration and misbranding of food and drugs in the US. The nauseating condition of the meat packing industry raised awareness to hazards in the marketplace, leading to a comprehensive food and drug law. Since 1879, nearly 100 bills have been introduced in Congress to regulate food and drugs, and in 1906, President Roosevelt signed the Food and Drugs Act. USDA Bureau of Chemistry was responsible for examining food and drugs for such "adulteration" or "misbranding". In 1927, the Bureau of Chemistry's regulatory powers were reorganized under a new USDA body, the Food, Drug, and Insecticide Organization, which was shortened to the Food and Drug Administration (FDA) three years later. One of the FDA's greatest and influential regulations is on pharmaceuticals, which includes drug approvals, safety, emergency preparedness, and many more for the production of drugs on the market.

Center for Drug Evaluation and Research (CDER)

The mission of FDA's Center for Drug Evaluation and Research (CDER) is to ensure that drugs marketed are safe and effective. Drug companies seeking to sell a drug in the United States must first test them in the laboratory. The company then sends CDER the evidence from these tests to prove the drug is safe and effective for its intended use. A team of CDER physicians, statisticians, chemists, pharmacologists, and other scientists review the company's data and proposed labeling. If this independent and unbiased review establishes that a drug's health benefits outweigh its known risks, the drug is approved for sale. An important drug safety issue is one that has the potential to alter the benefit/risk analysis for a drug in such a way as to affect decisions about prescribing or taking the drug. After drug approval, the FDA may learn new, or more frequent, serious adverse drug experiences from post-approval clinical studies or from clinical use. As new information related to a marketed drug becomes available, the FDA reviews the data and evaluates whether there is a potential drug safety concern. When a potential drug safety concern arises, relevant scientific experts within the FDA engage in a prompt review and analysis of available data to determine whether regulatory action is warranted.

World Health Organization (WHO)

Another regulation agency concerned with global public health is the World Health Organization (WHO). During the 1945 United Nations Conference on International Organization, Szeming Sze (Figure 5.4), a delegate from China, proposed to create an international health organization under the auspices of the, then new, United Nations. The constitution of the World Health Organization was signed by all 51 countries of the United Nations, and by 10 other countries, on 22 July 1946. Nowadays, the work of the Expert Committee on Specifications for Pharmaceutical Preparations (ECSPP), one of the WHO Expert Committees, consist of regulations on pharmacopoeia standards, quality control, production and inspections, distribution and good pharmacy practice.

Figure 5.4: Chinese diplomat Szeming Sze (1908 - 1998) co-founder who helped build the World Health Organization (WHO) in 1948.

From research laboratory to the patient

Once a pharmaceutical or biopharmaceutical company has isolated a compound, they believe to be a novel and effective drug, it must undergo rigorous federal testing to confirm its compliance with national standards. Since its inception the FDA have altered and improved many of its testing and requirements based on technical advancements and health concerns. Of its many federal regulations, the FDA and its most current requirements enforce not only drug purity but also, through the Current Good Manufacturing Practice (CGMP), well-maintained equipment, clear labeling with detailed identification and storage, and test facilities with adequate space to prevent contamination. These practices will ensure proper monitoring and control of facilities during research and production. Initially, a compliant company itself conducts multi-stage tests in the laboratory as well as in clinical trials for its intended drug. Sufficient data is collected to prove that a new or improved drug has been created for the betterment of personal health. New drug applications are completed to formally introduce a proposal for the US FDA. For government approval (Figure 5.6), all the independent company's research and data are sent to the CDER where numerous scientists review the data and propose labeling.

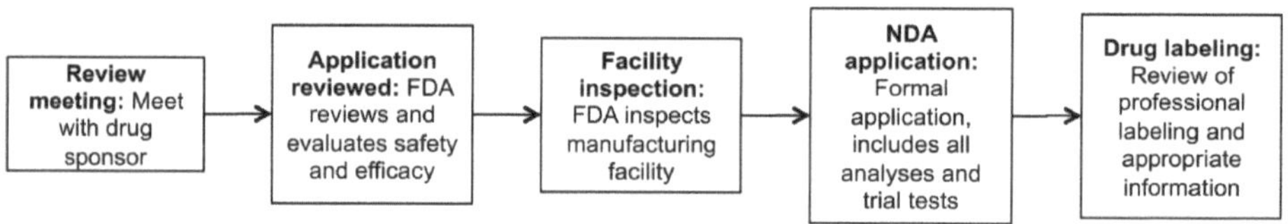

Figure 5.6: Diagram of the approval requirements for perspective drugs in the USA.

The Center for Drug Evaluation and Research is the ultimate body that reviews applications for pharmaceuticals and determine the appropriate medications to be marketed. Multiple divisions control different processes of drug approval:

- the Office of New Drugs Oversee Clinical Trials,
- the Office of Pharmaceutical Quality Test Drugs and Biotechnology for Safety,
- the Office of Medical and Regulatory Policy Develop and Review Guidelines,
- the Office of Surveillance and Epidemiology Monitor Post-Marketing Drug Usage.

For analytical chemistry purposes, the Office of Pharmaceutical Quality conducts rigorous chemical analysis to determine a drug's compliance to specific standards. FDA labs will test drugs based on such standards as determined by the United States Pharmacopeia (USP), a government nonprofit organization that creates and compiles quality standards for medicine and foods. All prescription, over the counter and animal drugs are required to conform to USP standards. The FDA and CDER will compare its trial drugs to the USP's standards of identity, strength, quality, and purity. They will even require the same quality standards for new and generic drugs by looking at impurities and dissolutions. After review and the drug's benefits are proven, the drug can be approved for the market ((Figure 11.8)).

For the development of analytical procedures to test characteristics of a drug, all methods are included in drug applications. A full description of manufacturing processes is used to confirm standards of identity, quality, and purity. If any FDA recognized procedures are applied, such a procedure is already documented in the United States Pharmacopeia/National Formulary. Analytical instrumentation and methodology are selected based upon specificity, limits of detection and quantitation, and accuracy and precision. Essential information is included pertinent to the analytical procedure: description of the basic tests and technology (i.e., separation or detection), equipment, operating parameters with optimal settings and ranges, reagents and sample preparation, and procedure. All mathematical transformations and data reporting consistent with an instrument's capabilities as well as reference blanks and drug performance and selectivity are included. Specifically, for chromatographic methods, retention times for comparison with reference standards must also be included. Statistical analysis of data utilizes techniques such as histograms, probability plots, and correlation coefficients to assess validity of data. All such previously conducted research by the pharmaceutical company is sent to the FDA and CDER for further tests.

Of the many of tests to determine efficacy and purity, a multitude of analytical chemical techniques can be used to isolate and analyze drugs. High performance thin layer chromatography (HPTLC) proves extremely versatile to analyze a variety of drugs by separating non-volatile mixtures. In high-performance liquid chromatography (HPLC) with a UV detector to monitor an elution wavelength spectrum proves powerful because it serves as both a quantitative and qualitative technique. The results collected can determine the compound present as well as the actual amount. Similarly, the degree of bioavailability can be evaluated to compare multiple doses or multiple drugs.

With an emerging technological industry, pharmaceutical nanotechnology is rising in its collection of drug-safety data. Nanotechnology, like nano-emulsions, can suspend or dissolve drugs for further analyses. One of the most powerful techniques for pharmaceutical analysis is liquid chromatography-mass spectrometry (LC-MS) because of its sensitivity in chromatographic separation coupled with a specific mass spectrometric detection. Drug metabolism studies, identification of impurities, and products from degradation are some of the major applications of LC-MS. Many of these techniques provide resources for independent pharmaceutical companies and the FDA/CDER to determine how biologically and chemically effective a novel drug may be. Ultimately if the results determine a high purity and bioavailability, a novel drug can begin production for the market.

Despite the USP's specific standards, the FDA ultimately maintains enforcement. The FDA's current domain oversees prescription and over-the-counter drugs, medical devices, food, tobacco safety, and many other health products. Within the food industry, the FDA focuses on quality of the food provided and its labeling, regulation nutrition charts and the additives and substances included. Failure to abide by the FDA's regulations can result in public defamation of a company through the press, ordering recalls on faulty products, petitioning a source to seize items in violation, and even criminal penalties on violators. Unique in the history the food and drug industry, the Dietary Supplements Health and Education Act (DSHEA) ruled to protect public health and ensuring quality of dietary supplements. These supplements are mainly unregulated and sold without federal regulated proof of safety or efficacy. Labels are leniently published under the pseudonym of Supplement Facts and restrict from claiming treatment or prevention of medical diseases. The FDA is only legally able to subject such supplements to regulations after proof of misbranding or harm results from such products. Ultimately, the FDA strives to protect and promote the public health industry by monitoring and controlling the drugs, foods, and products that can affect one's health. While there may be certain supplements exempt from such tight government scrutiny, the pharmaceutical industry focuses much of its analytical processing power to determine its products' safety and effectiveness for wide consumption.

A pharmaceutical company's ability to chemically analyze a drug is essential for the company to synthesize the drug correctly, and in accordance with standards defined by regulatory agencies of the industry. The techniques described above allow manufacturers to evaluate the composition of a drug from the earliest stages of development, to the packaging of the final product. This provides them with the information necessary to fine-tune protocols and maximize the efficiency of manufacturing processes. More importantly, such analytical chemistry techniques are required to address drug safety and efficacy concerns surrounding a pre-approved drug. Manufacturers must identify possible impurities and prove the drug is compliant with identity, strength, quality, and purity standards before it can be submitted for review. In order to detect impurities contained in a final product and prove that the drug meets certain regulatory criteria, pharmaceutical manufacturers must utilize chromatographic and spectrometric methods, among others, to collect the necessary data. These analytical methods, and the data that come from them, are instrumental in maintaining transparency between pharmaceutical manufacturers, regulatory agencies of the industry, and consumers.

As long as there are analytical methods available to indicate a drug's expected safety and efficacy on the market, regulatory agencies of the pharmaceutical industry will be able to oversee the development and commercialization of emerging medications and thereby protect consumers. They do so by establishing standards, derived from analytical methods, which pre-market drugs must meet, in order to apply for approval. Today, the modern approach to drug therapy is shifting as new types of medication are developed. The exciting development of personalized gene and nanoparticle-based therapies brings with it the challenge of regulating such medications.

Personalized medicine presents an opportunity for significant advances in the way pharmaceutical companies combat disease and will have to be regulated by an approach different than how commercially available drugs are regulated now. These new medications are created with chemical makeup specific to individual patients based on their genetics and the markers over or under expressed at the site of interest and are therefore different for virtually every patient. For that reason, performing the same analytical techniques and referencing the same standards set for commercial drug development would be an insufficient approach to determine the safety and efficacy of a personalized medication. Bearing this in mind, a new regulatory approach must be developed for such medications, involving more advanced tools and screening. Already, the FDA has deployed a number of programs designed to regulate the science in support of personalized medicine. The High-Performance Integrated Virtual Environment (HIVE) is a database utilized by the FDA, which stores and analyzes the genetic code of individuals in reference to the genomes of other individuals who have been treated with personalized medicine. This program, and many like it, illustrates the potential for more analytical techniques to surface in the future to regulate new medications.

Good practice (GxP)

Good practice laws were officially enacted in 1978 by the FDA and have been updated consistently over the past 40 years. Good practice laws cover all steps of development of the drug from laboratory, to clinical trials, manufacturing, and distribution. In the laboratory, GLP dictates quality standards for how a study is conducted, how to collect data, and how to report results. For example, multiple scientists must sign off on a certain method used for a synthetic step in the development of a drug in order to ensure there was no falsification of data, and that records indicate what actually happened. Another part of GLP is quality assurance (QA). A QA auditor who has to ensure that all the people conducting science experiments are qualified and confirm that all the methods were correctly followed ensures this. The goal of GLP is to ensure that good science is conducted.

Once a drug is taken to clinic it, GCP guidelines determine how a drug may be administered to humans safely. GCP guidelines come from a code of conduct from the Declaration of Helsinki in 1964 and were later combined with additional guidelines from the International Conference on Harmonization (ICH) in the 1990s. These guidelines were set in place to ensure humane treatment of human subjects in drug trials, specifically listing consent and safety of patients as critical to clinical trials. Part of GCP mandates that pharmaceutical companies must ensure efficacy, quality, and adverse effects of drugs prior to testing them in humans. In order to reach these goals, clinical trials are separated into multiple stages (Table 5.4). In these four phases companies are able to determine how much of a drug is safe for humans and eventually test novel drugs against the current available drugs. Breaching GCP can lead to massive wastage of money and resources for governments, health systems, and pharmaceutical companies.

Once a drug passes the clinic it can be produced on a large scale and be distributed widely. In order to ensure the same quality and consistency of drugs GMP and GDP the FDA enacts regulations. The goal of GMP and GDP is to take proactive measures to ensure that companies minimize contamination and errors, in turn protecting the consumer from harmful consumption of an impure or toxic drug. GMP and GDP regulations are intentionally vague allowing each individual company to decide how to regulate sanitation, personnel qualifications, and equipment functionality. Breaching GMP regulations can lead to large fines and sanctions against a company. Similar to GLP and GCP, quality is assured through audits.

Phase	Primary goal	Dose	Patient monitor	Typical number of participants
Preclinical	Testing of drug in non-human subjects, to gather efficacy, toxicity and pharmacokinetic information	Unrestricted	Scientific researcher	not applicable (*in vitro* and *in vivo* only)
Phase 0	Pharmacokinetics; particularly, oral bioavailability and half-life of the drug	Very small, sub-therapeutic	Clinical researcher	10 people
Phase I	Testing of drug on healthy volunteers for safety; involves testing multiple doses	Often subtherapeutic, but with ascending doses	Clinical researcher	20 - 100 normal healthy volunteers (or for cancer drugs, cancer patients)
Phase II	Testing of drug on patients to assess efficacy and side effects	Therapeutic dose	Clinical researcher	100 - 300 patients with specific diseases
Phase III	Testing of drug on patients to assess efficacy, effective- ness and safety	Therapeutic dose	Clinical researcher and personal physician	300 - 3,000 patients with specific diseases
Phase IV	Post-marketing surveillance, watching drug use in public	Therapeutic dose	Personal physician	Anyone seeking treatment from their physician

Table 5.4: Phases of pre-clinical and clinical trials involving new drugs.

Patents

Patents are generally important regulations in any chemical industry; however, they play a larger role in pharmaceuticals. Patents are the main incentive for companies to continue R&D and provide stability in the drug market. The US Patent office grants 20 years of exclusivity to an invention from the date of ling, with certain extensions or limitations based on the category. If the drug is novel, companies can apply to the FDA for 5-year extensions. In addition, if a drug is made to target a disease that affects young children, the drug can maintain exclusivity for up to 6 more months.

In recent times, patents have been seen as very controversial in the industry. They allow for great investment in research and allow for companies to make profits and keep producing. On average, it takes about $2.6 billion dollars of investment and 10 years to put a drug from research to market. Without patents, there would be little investment into research. Additionally, companies would wait until one company puts the drug on the market and just reproduce the same drug at lower prices and undercut the competition, essentially benefitting off another company's research without paying for R&D. For example, the company Amgen is valued at $7.72 billion dollars but has only produced two successful drugs: Neupogen, and Enbrel. Amgen has been able to pro t from its drugs' exclusivity in the last 20 years. On the other hand, without competition and no price limit, companies can overprice their drugs, leading to inaccessibility for large portions of the population.

Orphan Drug Act and Rare Disease Act

Another set of regulations that have influenced R&D in pharmaceutical companies is the Orphan Drug Act and the Rare Disease Act. Most of the R&D budget goes to development of drugs that affect a large population of people because that's where all the profits are for the companies. The problem is that many people who suffer from rare diseases get neglected and go untreated. To respond to this problem, the government has subsidized R&D into rare diseases through these two acts. This has increased R&D into pharmaceutical treatments for rare diseases, also known as orphan drugs. These acts have allowed several people to get treated for several diseases such as muscular dystrophy, ALS, Huntington's, and Parkinson's disease. In addition, to further incentivize pharmaceutical companies, the FDA has offered extensions of 7 years for orphan drugs, providing another huge area for pro t for the companies.

Innovation needs

With the industry wide reduction of R&D budgets, pharmaceutical companies need to innovate and find ways in order to reduce the time and cost of R&D while increasing returns on investments. In order to do this new technology must be used such as machine learning, genome sequencing, 3D printing, and nanotechnology. These technologies can revolutionize drug discovery and manufacturing and reduce the burden on analytical chemists.

One of the biggest problems for analytic chemists is the time it takes to characterize and separate the vast number of drugs that could potentially be a treatment. Many of these drugs show no human activity and waste the time of many scientists. By incorporating machine learning and artificial intelligence into drug discovery, better guesses can be made by scientists. For example, with a crystal structure of a protein, artificial intelligence will be able to make the optimal guess of a drug that can't into the binding pocket of a protein. This will guide chemists in the correct direction instead of making several drugs that are all guesses. This reduces the amount of work for analytical chemists and reduces the cost of R&D. In order to make this change, the pharmaceutical industry must increase recruitment of data scientists and technology employees.

Another innovation in pharmaceuticals is genome sequencing. This technology allows researchers to identify gene abnormalities or biomarkers in individuals to predict which type of patients will be most bene ted by a certain drug and which patients will have severe side effects. Recently Obama's $215 million Precision Medicine Initiative has spurred the growth of several genomics programs in industry. Companies such as GlaxoSmithKline, Roche, AstraZeneca, and Abbvie have already been working with the FDA to develop genomics programs in order to expedite the drug discovery process. Genome sequencing could ultimately cut time and costs in R&D and lead to higher incentives to conduct R&D.

One of the jobs of analytical chemists in pharmaceuticals is to examine the concentration and quality of active ingredients and assess how the drug will disperse throughout a human. Since every human is different it can be di cult to design a pill or tablet that works just as well in every human. 3D printing can revolutionize individualistic treatments by delivering specific doses of drugs with specific diffusion properties. 3D printing allows a pill to be made by stacking layers of the drug until the desired dosage is reached. Researchers at the School of Pharmacy of University College of London have analyzed how the surface area and volume contribute to a drug's diffusion in the body. With 3D printing, companies can vary the surface area to volume ratio for each pill and provide the best option for patients. This would be di cult to do with standard manufacturing practices. Aprecia Pharmaceuticals has utilized this technology and has gotten approval from the FDA for their epilepsy drug Spritam. The long-term goal of 3D printing is to have each patient be able to download and print personalized pills. This goal would eliminate the need for analytical chemists to create a general pill or tablet that has some general diffusion pattern and will allow analytical chemists to have more freedom with drug design.

Finally, pharmaceutical companies can look towards nanotechnology to reduce the time and costs of R&D. Currently many analytical and synthetic chemists must add excipients in addition to the active ingredient in order to improve diffusion across membranes or to increase solubility; however, with nanotechnology, scientists can guide the delivery of a drug to specific locations using infrared or ultraviolet radiation. This can improve drug activity and reduce the time it takes to design a drug thus improving profits for the companies.

Key analytical techniques

Table 5.5 lists the analytical techniques commonly used in the pharmaceutical industry.

Technique	Function of Technique
Liquid chromatography (LC) and high-performance liquid chromatography (HPLC)	Separate and identify the chemical components of a chemical mixture
Gas chromatography (GC)	Separate the mobile and stationary gas phases
Nuclear magnetic resonance spectroscopy (NMR)	Observe local magnetic fields around atomic nuclei to identify a sample
Mass Spectrometry (MS)	Uses electric and magnetic fields to measure the mass of charged particles
Amperometry	Detects ions in a solution based on electric current or changes in electric current, measures concentration
Polarography	Analyzes solutions of reducible or oxidizable substances
Capillary electrophoresis (CE)	Separates charged analytes through a small capillary under the impact of an electric field
Flow injection analysis (FIA)	Records changes in absorbance, electrode potential, or other physical properties resulting from the passage of the sample material through the flow cell.

Table 5.5: Analytical techniques used in the pharmaceutical industry.

Chromatography

Chromatography is used to separate mixtures of substances based on certain properties such as size, charge, or polarity. In chemistry, separation by polarity is most common. Differences in polarity are exploited to separate mixture of compounds to assess the components in a few different methods.

The simplest technique is known as thin layer chromatography (TLC). A thin layer of adsorbent, usually silica or alumina, is laid on top of a base surface, commonly glass or aluminum foil. The adsorbent is a strongly polar compound and interacts with other polar compounds. Sample of the product is spotted or dropped onto the plate at the bottom. The plate is then placed approximately 0.5 cm deep into a solvent mixture that is much less polar than the adsorbent. The solvent is then pulled up the plate using capillary action. As the solvent moves up the plate, more polar compounds interact strongly with the plate and do not travel, while the less polar compounds move with the solvent up the plate. The result is to separate a mixture of compounds by their polarity. The compounds on the plate are visualized using UV absorbance. If the compounds of interest are not UV active, various stains can be used to react with the compounds on the plate and cause them to change color for observation. TLC separations are quantified using R_f values, defined as the distance travelled by the compound of interest divided by the distance traveled by the solvent. TLC is a very valuable technique because it requires only a small amount of sample, has minimal equipment costs or preparation associated with it, and can give results rapidly. However, compounds that have similar polarities cannot be easily separated. TLC also has low sensitivity, so it cannot be reliably used to detect trace contaminants.

In order improve separation another technique called high performance thin layer chromatography (HPTLC) can be used. This technique uses higher quality plates with smaller adsorbent particle size. This allows the plate to give both better separation and higher sensitivity. In addition, the smaller particle size reduces the runtime of a TLC separation.

Another version of chromatography analysis commonly used is high performance liquid chromatography (HPLC). Liquid chromatography (LC) is a commonly used technique for purifying solutions by separating them by polarity and collecting samples individually. HPLC is to LC as HPTLC is to TLC. The additional advantages of HPLC make it useful to chemists as an analytical method. In HPLC, specialized equipment is used to push solvent, called the mobile phase, through a column, referred to as the stationary phase. This gives much better separation than what is seen in traditional liquid chromatography and allows chemists to see components of mixtures with superior resolution than TLC. Specific HPLC columns can be used to enhance separation even further, including chiral columns that can be used to separate enantiomers in a racemic mixture. The different components are also visualized using UV spectrometry, and if the compounds of interest are not UV active, then they cannot be detected. Since the solid phase is not exposed like it is in TLC, certain stains cannot be used.

There are many other forms of chromatography that are useful to chemists like gas chromatography which involves vaporizing a sample and passing it through a hollow column lined with a stationary phase (usually a thin layer of liquid or polymer). The solid phase uses the same polarity principle as seen in both TLC and HPLC. The sample is carried on the mobile phase, an inert gas, most commonly helium. The sample travels through the column where more polar compounds take longer to elute as they have more interactions with the solid phase. This causes the mixture to separate by polarity. At the end of the column is a detector, often a thermal conductivity detector (TCD), a flame ionization detector (FID), or both.

A thermal conductivity detector works by passing the sample around a tungsten-rhenium lament with a current traveling through it. The inert gas does not disturb the thermal conductivity or the metal, but analyte molecules can interact with the metal and decrease the thermal conductivity. A flame ionization detector works by pyrolyzing carbon-containing compounds to form cations and electrons, which generates a current for detection. Gas chromatography has an advantage because it does not rely on UV activity in order to detect the components of the solution. However, it does require the analyte to be volatile enough to be vaporized.

Mass spectroscopy

While chromatography is excellent for separating mixtures, it is unable to determine the components of a mixture. If the components are already known, it is possible to make educated guesses about which fractions belong to which component based on their relative polarities. If the components of the mixture are unknown, are not UV active, and/or cannot be stained, then it can be di cult to know what species you separated in your chromatography. This is where mass spectrometry comes in.

Mass spectrometry works by ionizing a sample in the gas phase and measuring the mass to charge ratio of the ions. There are multiple techniques for both the ionization step and the mass measurement step:

- Electron ionization (EI) is common because it is applicable to most organic compounds, with the only requirement being that they are volatile upon heating. Once the sample enters the gas phase, it is irritated with electrons where it acquires excess energy, causing it to fragment into ions. The mass of these fragments can then be measured and used to assess the mass of the sample as a whole.
- Electrospray ionization (ESI) is a technique where a solution of sample is ejected through a small tube into a strong electric field in a vacuum. This causes the solvent to dissolve such that the remaining sample particles pick up a charge. These ions are then passed into the analyzer to detect their mass to charge ratio. Because this technique causes samples to pick up multiple charges, it is unsuitable for smaller molecules. Thus, ESI is most commonly used for large molecules such as organometallics, polymers, and proteins.
- Matrix assisted laser desorption/ionization (MALDI) is a technique that relies on the use of a matrix, which absorbs the energy of a laser to excite the molecules into a gaseous ion state. Desorption is the process of ejecting the sample from the matrix, followed by moving the sample into the charged state. This technique is widely used for many types of samples. It is only limitation is an inability to detect below a certain molecular weight. This makes it unsuitable for use in small molecule chemistry.

Once the sample has ionized, it moves into the detector. The two most common detector types are the sector field mass analyzer and time-of-flight (TOF) analyzer. A sector field analyzer works by applying an electric field to the path of the ion, causing it to change course. This movement is more pronounced for lighter molecules allowing the machine to compare the relative mass to charge values. The time-of-flight analyzer functions similarly to a chromatography technique. All of the particles are accelerated into a path with the same potential to give them the same energy, so ion travel is dependent only on mass.

Mass spectrometry is a great compliment to both HPLC and GC. There are instruments are the market that combine the techniques, commonly referred to liquid chromatography-mass spectrometry (LC-MS) and gas chromatography (GC-MS). These instruments work by passing the sample first through the chromatography column to separate the mixture by polarity. As the different fractions come off of the column, they are passed directly into a mass spectrometry machine, most commonly MALDI-TOF. This means that the peaks given on the chromatography can be assigned a mass, making it easier to confidently identify the components of the mixture.

Capillary electrophoresis (CE)

Capillary electrophoresis (CE) is an electrophoretic technique based on the separation of charged analytes through a small capillary under the impact of an electric field. In CE, solutes are detected as peaks and the area of the peaks are proportional to concentration of solute. Compared to HPLC and gas chromatography, CE has a quicker time scale, uses low volume injections, utilizes cheap capillary columns, and often requires aqueous solutions. These beneficial characteristics make CE a valuable alternative to traditional separating methods. Flow injection analysis (FIA) is a recent attempt to automate the chemical analysis procedure. In FIA, a liquid sample is injected into a moving uninterrupted stream of carrier fluid. The injected sample forms a zone in that fluid and then moves to the detector. It utilizes a detector that records changes in absorbance, electrode potential, and other physical properties resulting from the passage of a sample material through the flow cell. Figure 5.7 depicts the different phases of FIA as a sample flows through the cell.

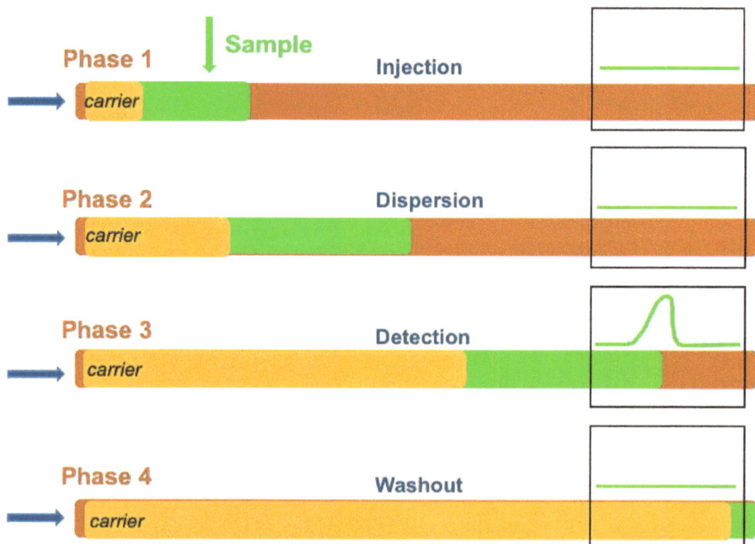

Figure 5.7: An illustration of the different phases of FIA. Reproduced from M. R. Siddiqui, Z. A. Al-Othman and N. Rahman, Analytical techniques in pharmaceutical analysis: A review. *Arab. J. Chem.*, 2017, 10, S1409. Copyright: Elsevier (2017).

Conclusions and future directions

Sound analytical chemistry is extremely important for the pharmaceutical industry. Due to shorter patents and smaller research budgets, analytical chemists have to find ways to be more economically efficient. The types of techniques that have the most utility in the industry are gas chromatography, HPLC, NMR, and mass spectroscopy. The advancement of capillary electrophoresis and flow injection analysis has made these techniques very important as well. These techniques are all essential for the drug discovery, manufacturing, and quality assurance processes that pharmaceutical companies go through to bring a drug to market.

The pharmaceutical industry will face a great deal of challenges in the future, but new technology will also provide more opportunities for growth. Big data and artificial intelligence may be able to help companies with the drug discovery process. Additionally, the possibility of 3D printing drugs could make manufacturing cheaper as well as making drugs more widespread across the globe. Analytical chemistry must continue to evolve along with the pharmaceutical industry while also cutting costs as the economics of the business become more complicated.

Bibliography

B. Aksu, A. Paradkar, M. de Matas, O. Ozer, T. Güneri, and P. York, Quality by design approach: application of artificial intelligence techniques of tablets manufactured by direct compression. *AAPS PharmSciTech*, 2012, **13**,1138.

R. Burbidge, M. Trotter, B. Buxton, and S. Holden, Drug design by machine learning: support vector machines for pharmaceutical data analysis. *Comput. Chem.*, 2001, **26**, 5.

Center for Drug Evaluation and Research. *Development & Approval Process (Drugs)*, https://www.fda.gov/Drugs/DevelopmentApprovalProcess/.

Center for Drug Evaluation and Research. *How Drugs are Developed and Approved*, https://www.fda.gov/drugs/developmentapprovalprocess/howdrugsaredevelopedandapproved/.

B. Dejaegher, HILIC methods in pharmaceutical analysis. *J. Sep. Sci.*, 2010, **33**, 698.

Z. Desta, N. Soukhova, A. M. Morocho, and D. A. Flockhart, J. Pharmacol. Exp. Ther., 2001, 298, 2508.

J. Ermer, Validation in pharmaceutical analysis. Part I: an integrated approach.*J. Pharm. Biomed. Anal.*, 2001, **24**, 755.

FDA, *Quality Systems Approach to Pharmaceutical Current Good Manufacturing Practice Regulations*, https://www.fda.gov/downloads/Drugs/Guidances/UCM070337.pdf.

FDA, *Patents and Exclusivity*, https://www.fda.gov/downloads/drugs/developmentapprovalprocess/smallbusinessassistance/ucm447307.pdf.

FDA, *Orphan Drug Act*, https://www.fda.gov/industry/designating-orphan-product-drugs-and-biological-products/orphan-drug-act-relevant-excerpts.

H. Gu, B. Hu, J. Li, S. Yang, J. Han, and H. Chen, Rapid analysis of aerosol drugs using nano extractive electrospray ionizationtandem mass spectrometry. *Analyst*, 2010, **135**, 1259.

A. Lavecchia, Machine-learning approaches in drug discovery: methods and applications.*Drug Discov. Today*, 2015, **20**, 318.

K. Mervatova, M. Polášek, and J. Martínez Calatayud, Recent applications of flow-injection and sequential-injection analysis techniques to chemiluminescence determination of pharmaceuticals. *J. Pharm. Biomed. Anal.*, 2007, **45**, 367.

L. Novakova, Nováková, L. Matysová, and P. Solich, Advantages of application of UPLC in pharmaceutical analysis. *Talanta*, 2006, **68**, 908.

L. A. Pastur-Romay, F. Cedrón, A. Pazos and A. Belén Porto-Pazos, Deep artificial neural networks and neuromorphic chips for big data analysis: pharmaceutical and bioinformatics applications. *Int. J. Mol. Sci.*, 2016, **17**, 1313.

S. D. Roughley and A. M. Jordan, The medicinal chemist's toolbox: an analysis of reactions used in the pursuit of drug candidates. *J. Med. Chem.*, 2011, **54**, 3451.

B. Saramento, The challenges of biopharmaceutical products for analytical requirements. *Pharm. Anal. Acta*, 2012, **3**, 1000e110.

A. Schumacher, O. Gassmann and M. Hinder, Changing R&D models in research-based pharmaceutical companies. *J. Transl. Med.*, 2016, **14**, 105.

M. R. Siddiqui, Z. A. Al-Othman and N. Rahman, Analytical techniques in pharmaceutical analysis: A review. *Arab. J. Chem.*, 2017, **10**, S1409.

S. Tian, J. Wang, Y. Li, X. Xu, and T. Hou, Drug-likeness analysis of traditional Chinese medicines: prediction of drug-likeness using machine learning approaches. *Mol. Pharm.*, 2012, **9**, 2875.

C. J. Van Boxtel, B. Santoso and I. R. Edwards, *Drug Benefits and Risks: International Textbook of Clinical Pharmacology*, Wiley, Hoboken, New Jersey (2001).

T. Vankeirsbilck, A. Vercauteren, W. Baeyens, G. Van der Weken, F. Verpoort, G. Vergote, and J. P. Remon, Applications of Raman spectroscopy in pharmaceutical analysis. *Trends Analyt. Chem.*, 2002, **21**, 869.

Chapter 6: Automotive Industry

Sreyas Menon, Raul Rondon, Pavan M. V. Raja
and Andrew R. Barron

Introduction

On 18th October 2015 Michael Horn, CEO of Volkswagen (VW) Group of America, testified before Congress admitting that the company purposely designed their cars to cheat on emissions tests. The defeat device in Volkswagen's diesel vehicles allowed alteration of their performance to conform to nitrogen oxides (NO_x) emissions standards when being tested. Software installed in Volkswagen's onboard computers (ECU) could detect when they were being tested under laboratory conditions as opposed to real world road conditions and would alter the engine mapping to curtail NO_X emissions under laboratory test conditions. Analytical chemistry was at the heart of both VW's cheating and the 2014 West Virginia University (WVU) study that revealed the fraud. The defeat device was a program that instructed vehicles to decrease the emission of NO_x by decreasing engine performance and by injecting excessive diesel exhaust fluid (DEF), containing urea (Figure 6.2), into the exhaust. The urea reduces the NO_x in the exhaust to produce CO_2, nitrogen gas, and water vapor.

Figure 6.1: Ex-CEO of Volkswagen (VW) Group of America Michael Horn (1961 -)

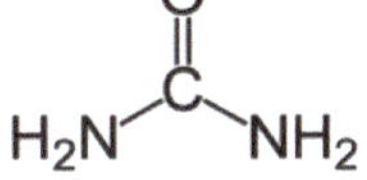

$$H_2N-\underset{\underset{\displaystyle NH_2}{}}{\overset{\overset{\displaystyle O}{\|}}{C}}$$

Figure 6.2: Structure of urea used as a diesel exhaust fluid (DEF) additive to lower NO_x emissions.

When WVU researchers tested Volkswagen's light duty diesel vehicles' emissions using portable emissions measurement systems, they found them to emit more than 20 times the EPA allowed level of NO_x. The portable emissions measurement system loaded onto the tested vehicles relied on common analytical chemistry techniques such as IR spectroscopy and chemiluminescence detection (CLD) to analyze the pollutants CO and NO_x. The use of these analytical techniques uncovered a multibillion-dollar scandal.

Volkswagen's developers were cheating to comply with stricter EPA standards in accordance with the Clean Air Act. Cheating allowed their vehicles to attain better fuel economy and road performance, while still appearing to comply with the EPA's NO_x standards when tested. The EPA sued Volkswagen and in 2016, the latter was ordered to pay $14.7 billion as settlement to their customers and for environmental damages. The Volkswagen scandal is a prime example of the impact that analytical chemistry can have, and the variety of clients for analytical chemists in the automotive industry, especially when it comes to analyzing emissions.

The automotive industry uses analytical techniques to test the performance of automobiles during development, manufacture, and operation. Automobile companies want to maximize fuel efficiency, responsive- ness, handling, and safety, in order to appeal to consumers. They also want to minimize costs, using less costly materials if possible while maintaining similar standards of performance and comfort. The regulatory sector is focused on ensuring that safety standards for vehicles are conformed to.

One of the most influential regulatory sectors that is involved with automobiles are environmental regulatory agencies. Led by the Environmental Protection Agency (EPA) in the United States, environmental regulatory agencies set policies and standards for all automobiles based on authority from legislation such as the Clean Air Act. In order to enforce these standards, the EPA and other agencies must translate these rules that are based on environmental and public health, into quantitative and testable scientific standards. Testing most of these standards involve devices based on analytical chemistry of varying complexity. Analytical chemistry has significant cross-disciplinary significance when determining the acceptable quantities of chemical toxins that vehicles can emit. Recently, a form of mass spectrometry has been applied to detecting black carbon soot and primary organic aerosols that are harmful to human health and are mostly a result of decomposition of lubricants rather than waste from the engine's exhaust.

The same principles and methods of analytical chemistry are used by materials industries that supply automotive companies. The use of analytical chemistry in automobiles is not restricted to the production process. Every automobile on public roads contains a plethora of chemical sensors that monitor their flown performance. Some of the most notable sensors in automobiles are O_2 sensors, which are both crucial in a car's self-monitoring system, and its compliance with environmental standards.

Overview

There are a lot of chemicals used in the manufacture and operation of automobiles, they are an important part of certain components such as air-conditioning, engine, battery, lubricating fluids, and emission control. Automobiles are equipped to monitor their flown performance. As an example, analytical chemistry is at the heart of the O_2 sensor (lambda sensor) that monitors the air-fuel equivalence ratio (denoted by λ). This sensor is crucial in ensuring the maximum efficiency of the engine.

The automobile has been often cited as a major cause of pollution and global warming. As part an effort to stop global warming, many countries set a permissible level for vehicle exhaust emissions. To resolve this issue, major automobile companies hire analytical chemists to control carbon emission of their products.

The primary pollutants in air include CO and oxides of nitrogen and sulfur. While these pollutants are produced naturally, their concentration levels in the atmosphere are exacerbated by emissions from vehicles. A major source of global warming is CO_2, and it is released by the combustion of fossil fuels in the internal combustion engine. To control carbon emission, analytical chemists quantify how much CO_2 is emitted from the exhaust using various techniques.

Particulate matter (PM) is another concerning category of pollutants found in automobile emissions. The particle range, from nanoparticles to those up to 2.5 μm in diameter and are a significant environmental health risk. Carbon soot particles are small, carcinogenic, and dynamic enough to enter human organs, and more recent studies indicates that they have a significant effect on climate change.

Particulate matter is commonly measured using a particle counter, which counts particles of a certain size. Detailed carbon soot particle analysis can be undertaken using soot particle aerosol mass spectrometer (SP-AMS). Nitrogen oxides (NO_x) (including, NO, NO_2, and N_2O) are also harmful pollutants produced by automobiles. to the World Health Organization, reducing NO_x emissions across the board could greatly improve human health. Oxides of nitrogen also combine with atmospheric oxygen to produce acid rain,

$$2NO_2 + \tfrac{1}{2}O_2 + H_2O \rightarrow 2HNO_3$$

SO_2 is another component of automobile (especially diesel) emissions that is also a precursor to that produces acid rain,

$$SO_2 + \tfrac{1}{2}O_2 + H_2O \rightarrow H_2SO_4$$

The harmful effects of acid rain illustrate the importance of detecting NO_x and SO_2 in automobile emissions.

Nitrogen oxides were the chemicals that were emitted by VW's light-duty diesel vehicles (model years 2009- 2015) in up to 40 times the quantity allowed by the EPA. In order to measure NO_x emissions, automotive companies and regulatory agencies must use a variety of devices, all of which are careful applications of techniques used by chemists. The most common technique is chemiluminescence of NO when reacted with ozone (O_3); however, new techniques are needed in response to tighter regulations and a demand for measuring emissions at the roadside rather than in a laboratory.

Key chemicals and materials

Table 6.1 gives some of the range of chemicals are used in the automotive industry, their source and estimated annual production. The volumes of chemicals listed in Table 6.1 are large and it can be assumed that there must be quite a large number of by-products of chemicals. Hence, it is important to detect those chemicals using various techniques.

Chemicals	Application	Annual production
1,1,1,2-tetrafluoroethane (Figure 6.3a)	Air-conditioning	175,000 tons
Ethylene glycol (Figure 6.3b), propylene glycol (Figure 6.3c)	Antifreeze	1.75 million tons
2,3,3,3-Tetrafluoropropene (Figure 6.3d)	Refrigerant	$800 million
Glycol-ether	Brake fluid	35500 tons
Duco	lacquer	5000 tons
Petroleum-based-hydrocarbon	Motor oil	542.71 billion liters (US consumption)
Methanol (CH_3OH)	Windshield washer fluid	110 million tons
Lithium cobalt oxide ($LiCoO_2$)	Lithium ion battery	124,000 tons
Zirconium dioxide (ZrO_2)	O_2 sensors	

Table 6.1: Selected commons chemicals used in the automotive industry.

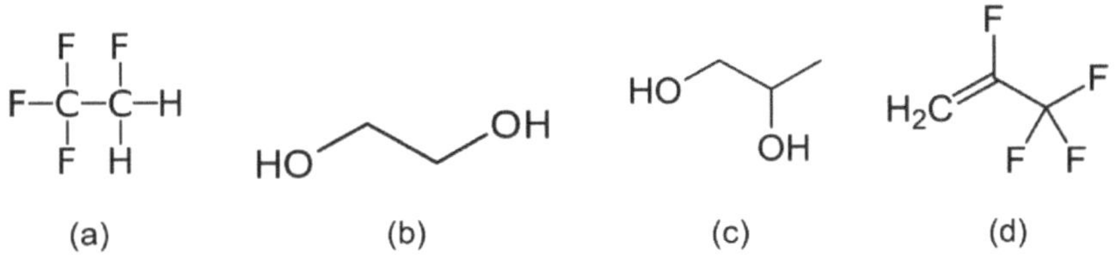

Figure 6.3: The chemical structure of (a) 1,1,1,2-tetrafluoroethane, (b) ethylene glycol, (c) propylene glycol, and (d) 2,3,3,3-tetrafluoropropene.

Challenges

There are many challenges that the automotive industry must meet, the most prevalent of which are the environmental challenges faced. In the early 1970, the rise of environmentalism brought out public doubts about the benefits of human activity for the planet. Curiosity about climate turned into anxious concern. Since then, researchers have found that the level of CO_2 in the atmosphere has doubled, and CO_2 is the major cause of climate change. With increasing worldwide demand for carbon emission control, the United States, ranked the second biggest CO_2 emitter globally, enacted the Clean Air Act (CAA) to regulate carbon emission. California has the right to adopt its flown emission regulations, which are often more stringent than the federal rules.

To fulfill the standard, major automobile companies seek various methods to minimize carbon emission. Internally, they improve engine efficiency to lower the incomplete combustion in the engine. Analytical chemists use various techniques to quantify exact amounts of particles emitted in the exhaust, and then they analyze what components comprise of those particles.

Regulations

Gasoline regulations are an area in which analytical chemistry plays a large role. Due to the developments of analytical methods governments have imposed legislations at federal and state levels. In the US, one of the more notable regulations in the automobile industry on the federal level is the Clean Air Act. This requires the EPA to monitor and regulate various aspects of automobiles such as the fuels they utilize, potential pollution to water, and emission products they might release. The EPA has put into effect different fuel standards such as limiting the amount of sulfur that is allowed in diesel fuel, constraining the amount allowable ethanol in fuel to 15 vol%, and setting the allowable gas volatility and Reid vapor pressure (vapor pressure measured at 37.8 °C) according to seasonal changes. Similarly, the EPA has regulated the concentration of different chemicals in automobile emissions. The EPA regularly monitors pollutants such as CO, NO_2 and particulate matter, as well as other air toxics, which are hazardous to the environment. Under the Clean Air Act, which is continuously updated, stricter and stricter regulations are imposed to limit the amount of these hazards which are produced from newer automobiles.

While federal regulations govern over all of the states, individual legislation has been put into effect by local government to further regulate fuel on a state level. Often times States will impose more rigorous regulations in order to meet certain air quality needs. For example, the Texas Natural Resource Conservation Commission (TNRCC) lowered the Reid vapor pressure regulation to avoid vapor lock that often occurs with high vapor pressure fuels in high ambient heat. In another example, in 2000 the TNRCC put into effect regulations to quell the increasing use of methyl *tert*-butyl ether (MTBE, Figure 6.4) in gasoline during the summer. Similarly, in 2009 Texas required that 2 vol% of diesel must be biodiesel, of which renewable diesel could not meet more than 25% of the requirement. Generally, regulations are becoming stricter to create a more-green future. On the other hand, in a state of emergency regulations are sometimes removed. For example, on 25th August 2017 the EPA waived its reformulated gasoline (RFG) regulation for Houston, which allowed Houstonians to utilize gasoline that produces higher levels of smog and toxic pollutants. The reason for this during this time Houston had been hit by a hurricane and was having trouble accessing RFG so the EPA dropped the national regulation for the State of Texas while they endured the natural disaster.

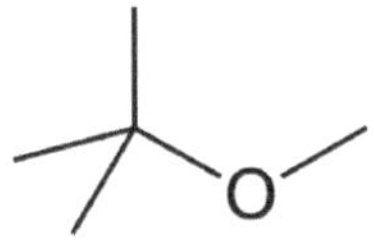

Figure 6.4: The chemical structure of methyl *tert*-butyl ether (MTBE).

While it may seem that the US heavily regulates its fuel and emission standards, its standards are quite lenient when compared with those observed on a global scale. For instance, taking CO_2 emissions for light vehicles (Table 6.2) that the current and future US standards are far less regulated than that of the EU shows the US lags behind in tackling carbon emissions.

Year of regulation	USA	EU
2015	155 gCO₂/km	130 gCO₂/km
2025	101 gCO₂/km	95 gCO₂/km

Table 6.2: USA and EU regulations for CO_2 (g) emitted per km distance travelled.

The EU along with many other countries have put into effect carbon taxes to help reduce overall emissions. Employing a carbon tax reduces emissions twofold, for one it increases overall cost of energy, which reduces overall demand, and secondly it increases costs of carbon-based fuels significantly more than that of gas and renewable energy to promote a switch to cleaner energy. The US's inadequacy in carbon emission regulations in automotive and all sectors has led them to be one of the world's largest producers of CO_2. In Figure 6.5 the CO_2 emissions of various countries in 2015 are displayed, showing the United States to historically and undoubtedly be one of the largest producers of CO_2, especially per capita (Figure 6.6).

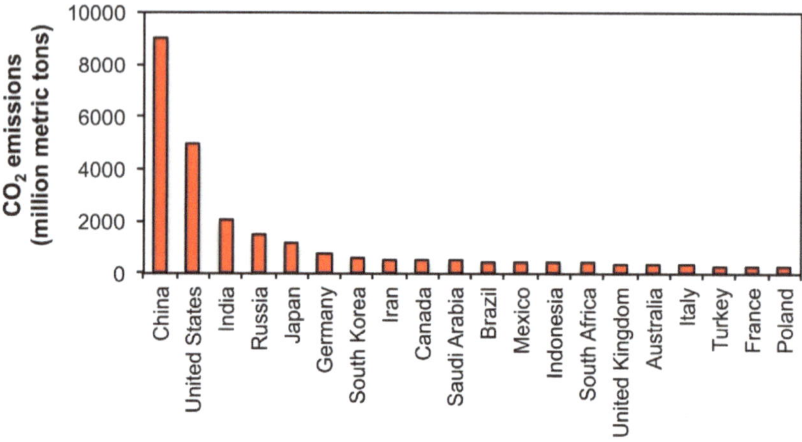

Figure 6.5: Total carbon dioxide emissions (million metric tons) from top 20 countries in 2015.

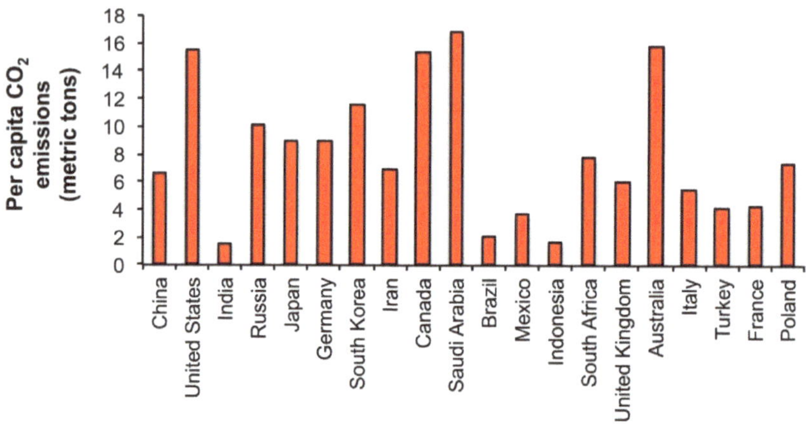

Figure 6.6: Carbon dioxide emissions per capita from top 20 CO₂ emitting countries in 2015.

Automobile emissions regulations are important globally because the CO_2 produced in one country influences the global population as a whole. With increased CO_2 levels, the climate will continue to drastically change. That being said it is unfortunate to see that the countries with the strictest regulations, often time doing the least harm to humanity are still suffering at the expense of other countries whose regulations have not evolved to keep pace with the changing world. These regulations, some of which have been in effect for decades, are continually revised as newer analytical methods become available to better detect smaller quantities of chemicals. The regulations in the automobile industry extend well beyond the regulations of chemicals in fuel, and the allowable chemical emissions produced by automobiles. But nevertheless, as humanity realizes the damage they are causing to the earth and newer technology arises, the effect will be the same. With more accurate instrumentation and better analytical methods stricter regulations will continue to be imposed to ensure the safety of humanity and the environment.

Innovation needs

Roadside emissions testing is a frontier for analytical chemistry in the automotive regulatory sector. Roadside emissions testing allows for the emissions of vehicles to be monitored without bulky equipment like an onboard system and also without the possibility of cheating or inaccuracy that arises in a laboratory measurement. Yet, roadside emissions testing is more challenging than in a laboratory or even onboard testing because it requires accounting for a variety of factors. One way in which roadside emissions are being test is through a technique called dilution tube sampling.

Dilution tube sampling works by collecting exhaust emissions, and then replicating the cooling and diluting of emissions in the actual atmosphere. By accounting for outside factors such as the cooling of exhaust emissions, temperature, humidity, volume and stream of atmospheric air, dilution tube sampling has become an effective and reliable way of emissions detection. In fact, northern regions of Virginia are offering state-run roadside emissions detectors that drivers simply drive through, and then are notified of their status of emissions, and then billed for the service.

NO_x emissions too are an important measurement concerning emission regulations of automobiles, as they are a harmful pollutant that contributes to the presences of hydroxyl radicals, secondary organic aerosols as well as acid rain. The standard method of detection of NO_x is by chemiluminescence, yet there are a multitude of disadvantages. One disadvantage is the interference that arises from other nitrogen containing compounds. Furthermore, in light of the VW emissions scandal and tightening regulations, government agencies are looking for more reliable ways to test for the levels of NO_x emitted. This calls for the introduction and enhancement of analytical techniques, and in the case of NO_x detection, that has come in the form of laser-based absorption spectroscopy.

Through the use of a continuous wave distributed-feedback quantum cascade laser and a continuous wave external cavity quantum cascade laser, there came about the ability to continuously and simultaneously measure both NO and NO_2. The two quantum cascade lasers converge and are both transmitted through a multi-pass glass cell, and detected by a single mid-infrared detector, where data was acquired and then analyzed by computer. By using reference cells containing high concentrations of NO and NO_2 and setting the laser to the frequency of absorption of the two compounds, the two lasers were able to continuously detect NO and NO_2 in sample.

Key analytical techniques

Table 6.3 summarizes various analytical techniques used within the automotive industry or for emissions monitoring.

Analytical technique	Specific method	Substances measured
Mass spectrometry (MS)	Soot particle aerosol mass spectrometer (SP-AMS)	Carbon particles from soot emissions
IR spectroscopy	Nondispersive infrared sensor (NDIR), Fourier transform infrared (FTIR)	CO, CO_2, NO_x
Chemiluminescence	Titration of NO with O_3; detection of emitted light with photomultiplier tube	NO_x
Particle counting	Electrical particle counting	Soot nanoparticles
Electrochemical analysis	Lambda sensor using a ZrO_2 element with platinum electrodes, or conductive TiO_2	O_2, air-fuel ratio (AFR)
Atomic spectrometry	Gas chromatography	NO_x, SO_2
Elemental analysis	CHNX analysis; measuring mass fraction of carbon, and nitrogen	Carbon and nitrogen
Thermal analysis	Catalytic combustion	Carbon monoxide

Table 6.3: A variety of types of analytical techniques are used to detect vehicle emissions.

In order to measure a variety of harmful emissions components, a range of analytical techniques are required. As shown in Table 6.3, mass spectrometry and particle counting are preferred techniques for microparticle and nanoparticle detection, while IR spectroscopy and chemiluminescence are the most common techniques for detecting small molecule pollutants such as CO_2 and NO_x. Other techniques, such as electro- chemical analysis used in small devices are crucial for automobile function. Less commonly used techniques include atomic spectrometry, elemental analysis, and thermal analysis, for these same pollutants.

Mass spectrometry

To measure particulate patter (PM) mass concentrations and chemical compositions in the atmosphere and in emissions, soot particle-aerosol mass spectrometry (SP-AMS) is used. This method can be applied not only to climate change research but also to combustion exhaust monitoring and source characterization. Most of all, it can be applied to carbon black containing particle quantification and analysis. The reason why carbon black is measured is that it is a pollutant emitted from gas and diesel engine, and other sources that burn fossil fuel. The carbon comprises a significant portion of PM (particulate matter), and it is a major cause of air pollution. Therefore, to quantify how much black carbon is released from the exhaust is important for preventing air pollution.

Soot particle-aerosol mass spectrometry has various advantages. First, SP-AMS vaporizes black carbon particles using intracavity laser absorption (1064 nm) and analyzed using electron impact ion source and high resolution, time-of-flight mass spectrometry. Second, laser vaporization is selective for black carbon and metal containing particles. Third, simultaneous operation with laser and standard tungsten vaporizer allows for detection and characterization of all major submicron ambient particle types (excluding uncoated dust and sea salt particles). The SP-AMS instrument is a standard HR-ToF-AMS with an added intracavity laser vaporizer. The SP-AMS, like other AMS instruments, provides average mass spectrometric measurements of the ensemble of sampled aerosols rather than measurements on a particle-by-particle basis.

Electrochemical O_2 sensors (lambda sensors)

The main purpose of an automobile's oxygen sensor is to monitor the engine's air-fuel ratio (AFR), and hence the efficiency of combustion of the fuel within the engine, as well as the performance of the vehicle's catalytic convertor. The AFR is defined by

$$AFR = m_{air}/m_{fuel}$$

where, m_{air} = mass of air in engine, m_{fuel} = mass of fuel in engine. A stoichiometric (ideal) air-fuel ratio is around 14.7:1.

Quantitative measurement of the exhaust O_2 concentration by the O_2 sensor relies on principles of electro- chemical analysis to compare an air reference sample with the exhaust across a sensor element. O_2 sensors are also called lambda sensors because they measure lambda (λ); a measure of how close the engine's air-fuel ratio is to the stoichiometric ratio. Vehicle oxygen sensors often use zirconium dioxide (ZrO_2) along with platinum electrodes to measure the partial pressure of O_2 electrochemically (Figure 6.7). The exhaust from the engine enters the sensor and comes in contact with the ZrO_2. Housed inside the device is the reference air, drawn from outside of the vehicle. If the concentration of O_2 (and thus partial pressure of O_2) is greater in the reference air, O^{2-} is conducted across the ZrO_2 material into the exhaust (Figure 6.8). This creates a positive potential as measured across a conductive Pt electrode. This potential is related by the Nernst equation:

$$E = E_0 - \frac{RT}{nF}\ln\left(\frac{a_{oxidized}}{a_{reduced}}\right)$$

In the case of O_2 sensors, this is applied as,

$$E = E_0 - \frac{RT}{nF}\ln\left(\frac{PO_{2,\,ex}}{PO_{2,\,ref}}\right)$$

where $PO_{2,ref}$ = partial pressure of O_2 in the reference air, $PO_{2,ex}$ = partial pressure of O_2 in the exhaust (sample), E = voltage generated, E_0 = standard reduction potential of O_2 to O^{2-}, R = gas constant, T = temperature, and F = Faraday's constant.

The exhaust analyzed by a lambda sensor can either be directly from the engine or after the exhaust has passed through the catalytic convertor. Lambda sensors are used upstream (in the engine) of the catalytic convertor to monitor engine performance and AFR in engine. The AFR in the engine is important for a number of reasons. One is that a proper air-fuel ratio results in the most fuel efficiency and highest performance of an engine.

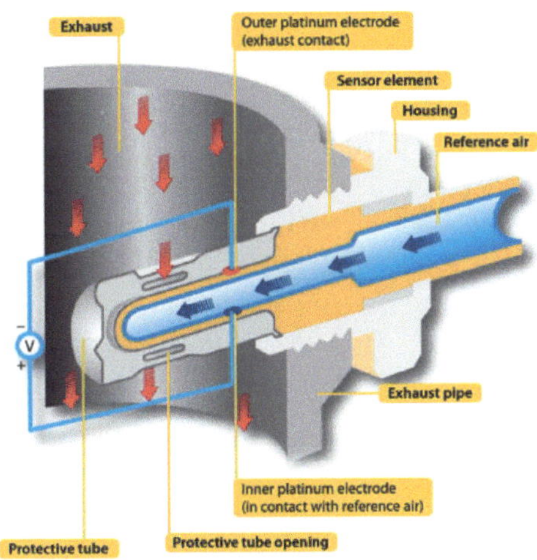

Figure 6.7: Cross section of a standard O_2 sensor using a ZrO_2 element and platinum electrodes. NGK Spark Plug Europe (https://www.ngk.de/en/products-technologies/lambda-sensors/lambda-sensor-technologies/zirconium-dioxide-lambda-sensor/).

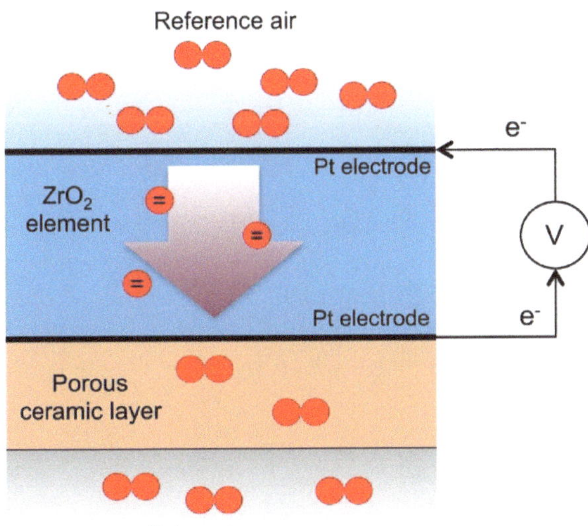

Figure 6.8: Flow of O^{2-} across the ZrO_2 element depends on exhaust O_2 concentration. A current in the Pt electrode is generated when O^{2-} is conducted through the ZrO_2 element.

A rich AFR (<14.7:1) indicates that the reaction mixture has excess hydrocarbon fuel. A lean AFR (>14.7:1) has excess O_2. Both rich and lean air-fuel ratios cause inefficiency (Figure 6.0). A rich AFR results in un-combusted fuel being emitted from the exhaust and an increase in black carbon soot particles in the exhaust due to incomplete combustion of the fuel. Excessive CO production is also the result of a rich mixture, as there is insufficient O_2 to produce O_2. A lean AFR results an increase in the emission of other environmental pollutants such as NO_x from the engine. Lean air mixtures result in higher temperatures, at which N_2 and O_2 react more readily to form NO_x.

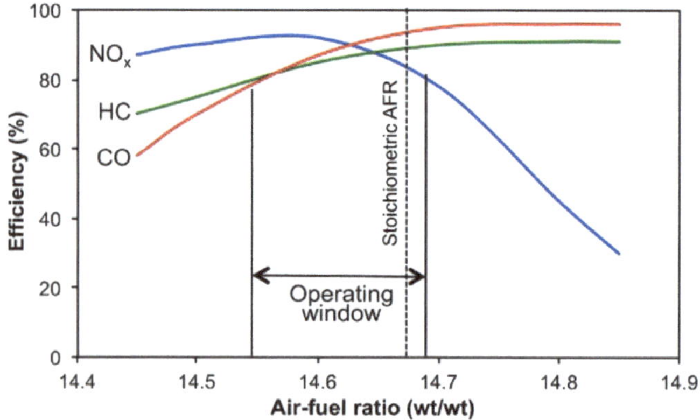

Figure 6.9: Idealized efficiency and emissions of three major pollutants as a function of air-fuel ratio.

A second O_2 sensor is also installed on all modern vehicles to monitor the performance of the vehicle's catalytic convertor (Figure 6.10). Comparison between the pre-catalytic and post-catalytic O_2 sensor voltage readings indicated the amount of O_2 reacted in the catalytic convertor. O_2 reacts with hydrocarbons to form relatively benign by-products such as CO_2 and water. If the catalytic convertor is not functioning properly, then the concentrations of O_2 detected by each O_2 sensor will be almost identical, and the voltages will be similar at each oxygen sensor. In modern vehicles, this similarity in voltage at each O_2 sensor will trigger the check engine light to appear on the driver's dashboard as an alert.

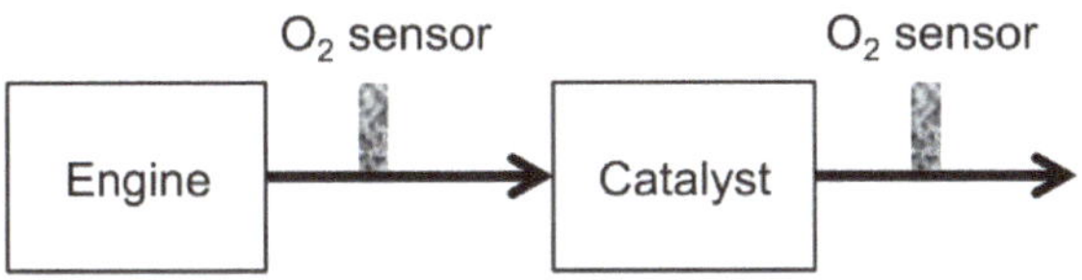

Figure 6.10: Design of dual O₂ sensor system to monitor performance of a modern automobile's catalytic convertor.

Chemiluminescence detection (CLD) NO_x analyzers

In order to measure NO (nitric oxide), gas samples from vehicle exhausts are mixed with ozone (O_3). The reaction of NO and O_3 produces light which can be measured when the photons emitted are detected by a photomultiplier in the reaction chamber. This is a kind of titration of the exhaust sample with O_3 to determine the concentration of NO.

Enhanced chemiluminescence analyzers have a NO_x converter (Figure 6.11). The NO_x converter consists of a hot catalyst (for example vitrified carbon) to convert all nitrogen oxides to NO, before sending the sample into the reaction chamber containing ozone. This allows the measurement of total NO_x (NO + NO_2 + N_2O_5) in the sample rather than just NO. These chemiluminescence detectors (CLD) are the standard in measuring NO_x, whether in the laboratory, in roadside testing, or in portable emissions measurement systems (PEMS).

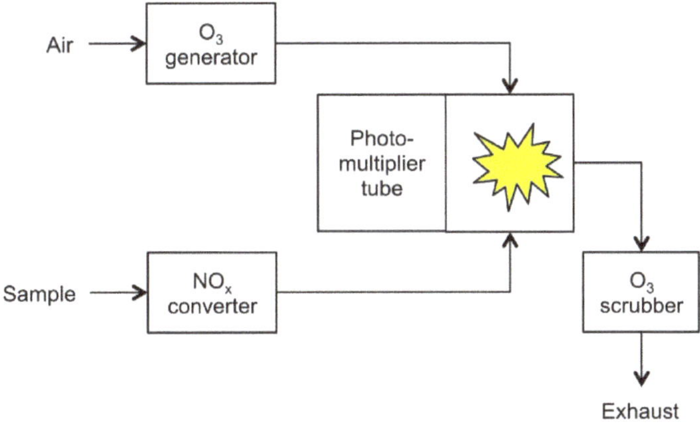

Figure 6.11: Schematic of standard NO_x chemiluminescence detector (CLD).

When measuring vehicle emissions, the sample comes from the vehicle's exhaust. An ozone generator to the reaction chamber/tube, which is connected to a photomultiplier tube, supplies O_3 (Figure 6.11). After passing through the reaction tube, the exhaust continues through the ozone scrubber so that the instrument does not release O_3 into the atmosphere.

Ozone supplied to the device reacts with nitrogen monoxide to produce a chemiluminescent excited state,

$$NO + O_3 \rightarrow NO_2 + O_2$$

that is quenched by any gaseous molecule present, and decays to ground state emitting light in the 600 - 875 nm range,

$$NO_2 \rightarrow NO_2 + h\nu$$

A photomultiplier tube (Figure 6.11) exposed to the sensor's reaction mixture then detects this visible light. A current response is generated in the photomultiplier tube, which can be measured. The current response (in amperes) is a linear function of the amount of NO present in the sample, with increasing NO leading to increasing current from the photomultiplier tube (Figure 6.12).

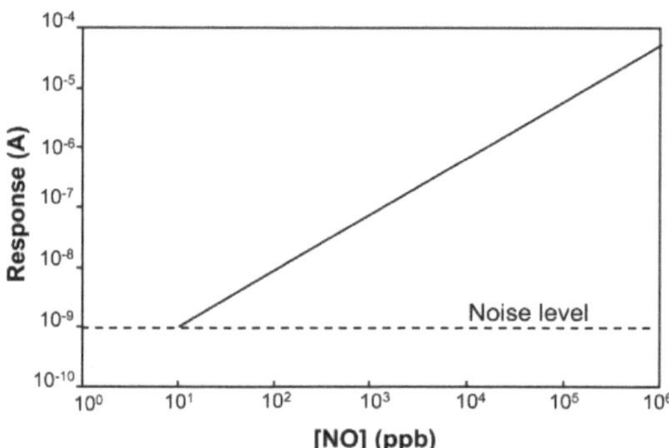

Figure 6.12: Ideal example of current produced by the photomultiplier tube as a function of NO concentration in the reaction chamber. Adapted from A. Fontijn, A. J. Sabadell, and R. J. Ronco, Homogeneous chemiluminescent measurement of nitric oxide with ozone. Implications for continuous selective monitoring of gaseous air pollutants. *Anal. Chem.*, **1970, 42, 575. Copyright: American Chemical Society (1970).**

Enhanced chemiluminescence detectors convert NO_2 to NO by passing the exhaust through high temperature catalysis before the exhaust reaches the ozone-containing chemiluminescence tube,

$$NO_2 \rightarrow NO + \tfrac{1}{2}O_2$$

The total concentration of NO detected after this reaction is then equal to the total concentration of NO_x ($NO + NO_2$) in the exhaust.

Non-dispersive infrared (NDIR) CO_2 sensors

Non-dispersive infrared (NDIR) sensors are the most commonly used tool to CO_2 concentration. NDIR spectroscopy is used in a variety of laboratory and portable emissions measurement systems. Non-dispersive infrared (NDIR) CO_2 sensors work by passing IR radiation through the exhaust sample (Figure 6.13). NDIR is considered nondispersive because unlike other types of IR spectroscopy, NDIR waves are not dispersed or deformed before passing through the sample. In practice, these sensors work the same as a benchtop spectrophotometer, relying on Beer-Lambert's law to calculate variations in gas concentration. The wavelength 4260 nm (2347.42 cm^{-1}) is used, since CO_2 is the only common gas to absorb at that wavelength. NDIR can also be used to detect other emissions such as SO_2.

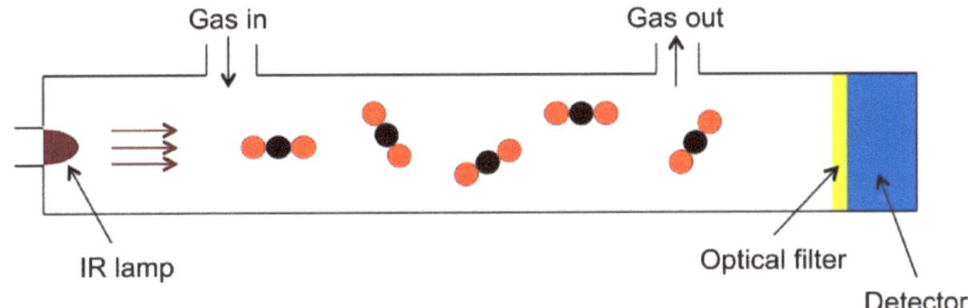

Figure 6.13: Schematic of a non-dispersive infrared (NDIR) CO_2 sensor.

Conclusion and future directions

Analytical chemistry is clearly a critically important aspect of the automotive industry, especially with respect to the detection of emissions. Yet the auto-motive industry faces a series of challenges, the most prevalent of which being state-mandated regulations on emissions. These state regulations vary globally, two of the most notable being the Clean Air Act put into effect by the United State, and the regulations mandated by the European Union. Interesting, although the federal government of the US has already put fairly strict regulations on emissions standards, state governments tend to take it even further, with even stricter regulations. Lastly, the EU and many other nations have emissions standards much more rigorous than those of the United States. While the US attempts to reduce its emissions, it still falls far behind in comparison to the intensity of the global effort to reduce emissions.

While these regulations pose serious challenges by which the automotive industry must concede, it also faces certain challenges to the field of analytical chemistry by which the field must fight to keep up with. As regulation standards become increasingly more demanding in both precision and reliability of emissions measurements, the field is pushed to produce innovative techniques to fit the increase in standards. One way in which this is being done is by roadside emission detectors, which are preferred in comparison with laboratory measurements as they are a more accurate depiction of how exhaust emissions actually impact the environment. Secondly is laser-based absorption spectroscopy, which provide a more precise and comprehensive measurement of NO_x emissions.

The future of analytical chemistry as is driven by the automotive industry seems limitless as state regulations only seem to be growing in demand for more precise and reliable instruments and techniques of detecting emissions from automobiles. This demand for a cleaner Earth itself is what drives the innovation of analytical chemical techniques to become even better and drive the field into the future.

Bibliography

Australian government Department of Environment and Energy, http://www.environment.gov.au/protection/publications/factsheet-sulfur-dioxide-so29.

T. C. Bond, S. J. Doherty, D. W. Fahey, P. M. Forster, T. Berntsen, B. J. DeAngelo, M. G. Flanner, S. Ghan, B. Kärcher, D. Koch, S. Kinne, Y. Kondo, P. K. Quinn, M. C. Saro m, M. G. Schultz, M. Schulz, C. Venkataraman, H. Zhang, S. Zhang, N. Bellouin, S. K. Guttikunda, P. K. Hopke, M. Z. Jacobson, J. W. Kaiser, Z. Klimont, U. Lohmann, J. P. Schwarz, D. Shindell, T. Storelvmo, S. G. Warren, and C. S. Zender, Bounding the role of black carbon in the climate system: A scientific assessment. *J. Geophys. Res. Atmos.*, 2013, **118**, 5380.

W. Dockery, Epidemiologic evidence of cardiovascular effects of particulate air pollution. *Environ. Health Persp.*, 2001, **109**, 483.

J. A. Driscoll, Acid rain demonstration: The formation of nitrogen oxides as a by-product of high-temperature flames in connection with internal combustion engines. *J. Chem. Edu.*, 1997, **74**, 1424.

EPA: *Air Quality*, https://www3.epa.gov/airquality/emissns.html.

EPA: *Federal Gasoline Regulations*, https://www.epa.gov/gasoline-standards/federal-gasoline-regulation.

EPA: *State Gasoline Standards*, https://www.epa.gov/gasoline-standards/state-gasoline-standards.

EPA: *Install Automated Air/Fuel Ratio Controls*, https://www.epa.gov/sites/production/les/2016-06/documents/auto-air-fuel-ratio.pdf.

EPA: *Learn About Volkswagen Violations*, https://www.epa.gov/vw/learn-about-volkswagen-violations.

R. J. Farrauto and R. M. Heck, Catalytic converters: state of the art and perspectives. *Catal. Today*, 1999, **51**, 351.

A. Fontijn, A. J. Sabadell, and R. J. Ronco, Homogeneous chemiluminescent measurement of nitric oxide with ozone. Implications for continuous selective monitoring of gaseous air pollutants. *Anal. Chem.*, 1970, **42**, 575.

D. Gibson and C. MacGregor, A novel solid state non-dispersive infrared CO_2 gas sensor compatible with wireless and portable deployment. *Sensor*, 2013, **13**, 7080.

G. Korotcenkov, *Handbook of Gas Sensor Materials: Properties, Advantages and Shortcomings for Applications Volume 1: Conventional Approaches*, Springer, New York (2013).

T. B. Onasch, A. Trimborn, E. C. Fortner, J. T. Jayne, G. L. Kok., L. R. Williams, P. Davidovits, and D. R. Worsnop, Soot particle aerosol mass spectrometer: development, validation, and initial application. *Aerosol Sci. Technol.*, 2012, **46**, 804.

Y. Otsuki, H. Nakamura, M. Arai, and M. Xu, The methodologies and instruments of vehicle particulate emission measurement for current and future legislative regulations. *Meas. Sci. Technol.*, 2015, **26**, 1.

Pennsylvania Department of Environmental Protection Inspection Manual (2000), http://www.elibrary.dep.state.pa.us/dsweb/GetRendition/Document-50339/html#bmk2.

T. D. Rapson and H. Dacres, Analytical techniques for measuring nitrous oxide. *Trends Analyt. Chem.*, 2013, **54**, 65.

Texas Natural Resource Conservation Commission, https://www.sunset.texas.gov/reviews-and-reports/agencies/texas-natural-resource-conservation-commission-tnrcc.

W. Wu, R. H.Chen, J. Y. Pu, and T. H. Lin, The influence of air–fuel ratio on engine performance and pollutant emission of an SI engine using ethanol–gasoline-blended fuels. *Atmos. Environ.*, 2004, **38**, 7093.

Chapter 7: Semiconductors and Nanotechnology

Angel Adrian Garces, Patrick Kelly, Tianyi (Emma) Wu,
Pavan M. V. Raja and Andrew R. Barron

Introduction

The fields of semiconductors and nanotechnology both rely heavily on analytical chemistry techniques to maintain the quality of the products and materials. Each of these fields requires very pure materials because of the significant changes in function that may result from tiny impurities. Analytical methods are used every day in these areas to ensure the purity of the materials and the functionality of the products.

Semiconductors

A semiconductor is a material that uses the excitation of electrons to conduct electricity. The advent of semiconductors began shortly after World War II in the 1940s due to funding for electronics and materials research development. The industry has grown over the past 70 years to become the fourth largest exporting industry in the United States. Worldwide sales of semiconductors reached \$335 billion in 2015 and are expected to continue rising. Currently, there are four broad semiconductor applications:

- logic for data manipulation,
- memory for information storage in computers,
- microprocessors for software tasks,
- converting analog to digital signals.

Energy levels in bulk materials are considered to be grouped into bands of infinitesimally spaced levels. Conduction is achieved by exciting electrons from the highest occupied band (valence band) to the lowest energy unoccupied band (conduction band), see Figure 7.1. Once in the conduction band, electrons are free to move throughout the material and thus conduct electricity. In metals, the conduction band overlaps the valence band, so tiny amounts of energy are required to move electrons throughout the material. As a result, metals are conductive. Insulators have a substantial energy difference between the valence band and the conduction band, and so electrons are typically not able to be excited by the conduction band. Semiconductors arise when the valence band and the conduction band are reasonably close in energy, such that electrons can move between the two bands with relatively small energy inputs (Figure 7.1).

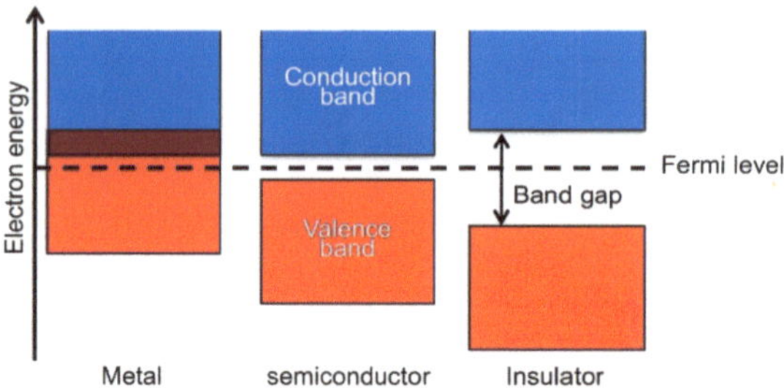

Figure 7.1: Schematic representation of the valence and conduction energy bands in metals, semiconductors, and insulators.

Typically, this energy comes in the form of thermal energy, though other forms of energy may be used. The difference between the valence band and the conduction band is known as the band gap. If the band gap in a material is small without altering the chemical composition of that material, the material is said to be an intrinsic semiconductor. However, semiconductors can be made by doping a pure material with another substance that introduces additional energy bands is called an extrinsic semiconductor. In one case, an additional unoccupied band can be added slightly above the valence band of the pure material (Figure 7.2). This type of doping creates a p-type semiconductor because the pre-existing valence electrons are more easily excited to the new conduction band. Alternatively, an additional occupied band can be introduced that is slightly lower in energy than the pre-existing conduction band. The new valence band electrons can be efficiently excited to the conduction band, thus creating an n-type semiconductor (Figure 7.2).

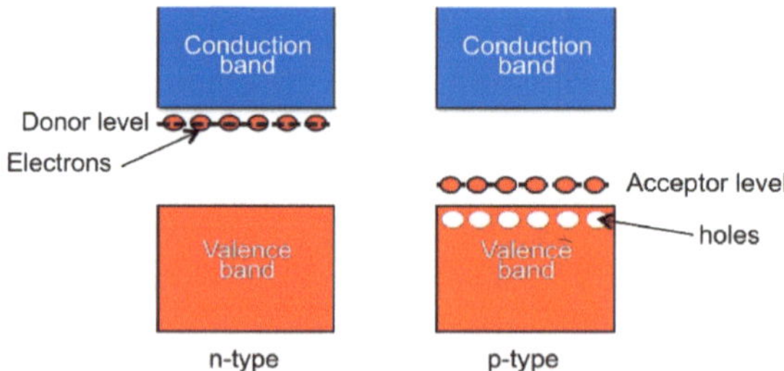

Figure 7.2: Schematic representation of n- and p-type extrinsic semiconductors (bands in doped semiconductors).

Semiconductors are made from a variety of materials, but the most common semiconductors are made from highly pure silicon doped with either a group III element such as aluminum (to create a p-type semiconductor) or a group V element such as phosphorus (to form an n-type semiconductor). Doping with an element that has fewer valence electrons than the bulk material (e.g., silicon doped with boron) essentially creates holes near the valence band, which creates a p-type semiconductor (Figure 7.2). Conversely, adding a dopant with more valence electrons than the bulk material (e.g., silicon doped with phosphorus) adds electrons near the conduction band and thus creates an n-type semiconductor (Figure 7.2). These two doping principles allow semiconductors to be made from a wide variety of materials.

Nanotechnology

Nanotechnology is a rapidly growing industry in analytical chemistry as the engineering of functional systems at a molecular scale. Due to the unique properties of nanomaterials, especially their surface-area-to-mass ratio and quantum effects, they are being used in a variety of industries. nanotechnology is used in the selective incorporation into high-end products, especially in the automotive and aerospace industries. In the energy industry, nanotechnology has helped the development of new catalysts for use in fuel cells. Batteries have nanostructured composite electrode materials, and thin films can be used within solar cells.

The change of the physical properties of nanoparticles as compared to bulk homologues is a result of an increased surface area to volume ratio. Their high surface area can improve the chemical conduction and performance of catalysis and structures such as electrodes (fuel cells and batteries). Other unique properties as a result of the high surface area of nanoparticles include increased strength and heat resistance. Because nanoparticles have dimensions below the critical wavelength of light, they obtain a level of transparency useful for packaging, cosmetics, and coatings. There are many types of nanoparticles, including metal oxide ceramic, metal, and silicate.

Key analytical techniques

Many analytical techniques are needed for semiconductors and nanotechnology to detect minuscule amounts of impurities and to visualize structures on the nanoscale. Table 7.1 lists some of these methods, what they are used for, and some of the limitations. All of these techniques are used in the semiconductor and electronics industry to guarantee the purity of the products.

Technique	Result	Detection limit	Limitations	Typical application
Transmission electron microscopy (TEM)	Images of nanoscale objects	~45 pm	Cannot see color, artifacts may arise from sample prep	Determining size of nanoparticles
Activation analysis	Concentrations of trace impurities	<1 ppb	Only allows 4-6 elements to be analyzed	Semiconductor quality control
Emission spectrography (flame test, plasma coupled, fluorescence emission)	Elemental analysis (concentration of each element)	~1 ng/mL (depends on element)	Destructive technique	Determination of alkali metal con- centration in pharmaceuticals
Electron paramagnetic resonance spectroscopy (EPR)	Environment around impurities with unpaired spins	<1 ppm	Sample must have unpaired electrons; electron spins are excited by a magnetic field	Determine whether Mn impurities incorporated into semiconductor nanocrystal or adsorbed onto surface
Energy dispersive X-ray analysis (EDS)	Trace impurities of a sample	~1000 ppm	Destructive technique	Semiconductor quality control
Extended X- Ray absorption fine structure spectroscopy (EXAFS)	Microenvironment of impurity atom (type of doping)	~1 ppm	Doesn't work as well for amorphous/ liquid samples	Detection of small concentrations of elements in the environment
Secondary ion mass spectrometry (SIMS)	Concentration of impurities in a sample	~1 ppm	Destructive technique	Semiconductor quality control
Rutherford backscattering spectrometry (RBS)	Concentration of impurities in a sample	$\sim10^{11}$ - 10^{15} atoms/cm^2	Only images in 2D	Analysis of trace impurities of thin films
Atomic force microscopy (AFM)	Images of nanoscale materials	0.5 nm	Requires air/liquid environment	Visualization of nanomaterials on a surface

Table 7.1: Key analytical techniques and applications.

Needs and challenges of the industry

The traditional semiconductor industry is limited in the ultimate smallest size of the materials that can be made because of the small concentration of dopant in a typical semiconductor. Because the concentration is so low (on the order of 10^{12} - 10^{13} atoms/cm^3) measuring the amount of dopant in a sample becomes very difficult as the size of the sample becomes small. Existing techniques have detection limits typically in the range of one part per million, but there require new methods to detect more minor quantities of impurities. The limit of detection thus limits how small semiconductor devices can be made reproducibly. There is, however, a potential use for semiconductors on the nanoscale especially in portable electronics.

Heat dissipation issues and bounds on purity levels may limit how small semiconductors can potentially be. This need to make semiconductors smaller will halt the shrinking of electronics until there appeared new technological advances. One potential way that researchers are attempting to overcome this is by making semiconductor wafers out of nanomaterials. Small wafers like these could be the future of transistors in electronics because they could potentially make processors smaller and more efficient. Researchers are also investigating other materials that may prove useful in electronics.

In addition to these challenges, the end of life challenge of semiconductors has not been widely explored. Environmental effects of used semiconductor devices have not been extensively researched despite the widespread use of semiconductors in technology. For continuous growth of the industry, more work needs to be done investigating the effects of semiconductors on the environment and protocols for disposal must be developed.

Challenges of nanotechnology mainly evolve around human health, environmental safety, and sustainability. Nanoscale materials often behave differently than those with a more massive structure. For example, a bulk aluminum product is perfectly safe whereas nano-sized aluminum is highly explosive. Because nanoparticles often have a higher toxicity, as compared to bulk analogues, hazards from engineered nanoparticles, resulting in tissue damage, inflammatory reactions, cardiovascular and other extrapulmonary effects, require awareness and safety precautions. Nanoparticle discharge into the environment needs environmental risk assessment and prevention as well. The source of this kind of discharge can come from production/transport and storage of intermediate and finished products, industrial waste, the release of particles during use of the products, as well as diffusion, transport and transformation in the air, soil, and water.

Regulations

Semiconductor industry

For semiconductors to perform their intended function, they must be extremely pure. In the synthesis of silicon semiconductors, silicon crystals are purified to 99.9999% purity. Once the required purity is obtained, chemical-mechanical polishing polishes the silicon wafers smooth. This process destroys any imperfections on the wafer disk with harsh corrosive chemicals. After eliminating flaws, the wafers are coated with photoresists and washed repeatedly. All of these processes are carried out in semiconductor fabrication facilities, or fabs, which are buildings with rooms that are completely dust-free. The air that goes into fabs is filtered to remove any impurities that could touch the wafers. These filters remove any impure compounds down to the width of 1% of a human hair. All workers must be covered from head to toe in bunny suits so that they do not introduce any contaminants to the room. The rigorous sterile technique is employed to protect the wafers from contaminants that could compromise their function.

Many chemicals required to produce functioning silicon wafers are harmful to humans. Since the semi- conductor industry only developed 70 years ago, the process of developing regulations for chemical exposure and environmental impact is ongoing. The need for chemical controls highlighted after a lawsuit in 1996 against IBM by former employees claiming that exposure to hazardous chemicals during work hours contributed heavily to the death of several employees was led. The trial resulted in a study by the United States Health and Safety Executive to determine whether the levels of exposure to carcinogenic chemicals in wafer fabrication facilities were at an unhealthy level.

Semiconductor facilities create a substantial amount of dangerous waste due to the harsh chemicals required for processing. Approximately 3787 gallons of contaminated water are generated in the production of a silicon wafer. This causes problems with drinking water contamination, specifically with trichloroethane, in California during the 1980s. More recently, the semiconductor industry has attempted to focus on cutting down emission of harmful CO_2 based and fluorinated gases.

According to the EPA, three classes of fluorinated gases (F-gases) contribute to the acceleration of climate change: hydrofluorocarbons (HFCs), perfluorocarbons (PFCs), and sulfur hexafluoride (SF_6, Figure 7.3). F-gases are present in refrigerators, air conditioning units, and in some aerosols. HFCs have been shown to deplete the ozone layer, but not nearly as much as CO_2. However, HFC contributes much more heavily to global warming than CO_2, which means that even small amounts of HFCs in the atmosphere can add significantly to global climate change. PFCs have similar effects to HFC regarding global warming. Sulfur hexafluoride is especially dangerous to the environment since it undergoes decomposition to produce oxy fluorides. Limiting the production of greenhouse gases in the semiconductor industry while searching for alternative methods that do not create copious amounts of greenhouse gases is an ongoing process.

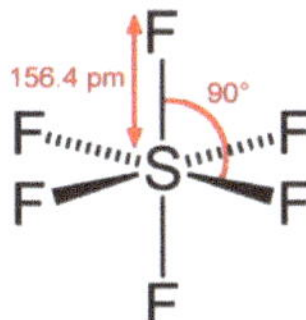

Figure 11.19: Structure of sulfur hexafluoride, SF_6.

Nanotechnology

Specific regulatory measures have been implemented to minimize environmental and health risks of nanotechnology. For example, European Commission-funded research projects implemented NanoSafe and NanoSafe2 to study risk in production and use of nanoparticles and NanoDerm to investigate the quality of skin as a barrier to ultra- ne particles. The Organization for Economic Cooperation and Development (OECD), which includes most of the industrialized nations of the EU has taken initiatives to test 14 generic nanomaterials for health and environmental effects while establishing a database for sharing research information on potential adverse effects of manufactured nanomaterials. Another example is UK's government's agenda on nanotechnologies, which includes setting up a research coordination group to investigate risks from nanoparticles, and risk assessment associated with medicines and medical devices. Additional measures include the safety of unbound nanoparticles in cosmetics and other consumer products, the disclosure of testing methodologies used by industry, labeling requirements on consumer products, and sectoring specific regulations for products of nanotechnologies, in additional to REACH a European level.

There are currently no regulations governing nanotechnology in the United States, although the California EPA has begun to request analytical data from manufacturers of carbon nanotubes to study the environmental implications. Part of the regulatory challenge remains regarding gaps in the oversight system. For example, certain product types that use nanotechnology and involve high human exposure, such as cosmetics and dietary supplements, are subject to laws that prohibit effective oversight.

Conclusions and future directions

Nanotechnology will play a significant role in a variety of industries, ranging from medicine to electronics. With more comprehensive risk assessment and preventative measures to ensure human health and environ- mental safety, future areas of nanotechnology such as electronics and communications chemicals and materials, manufacturing, energy technologies, space exploration, environment and national security. Regarding medical use, improvements in drug delivery, repair/replacement, hearing, and vision can impact in the future. In the food and agriculture market, nanoproducts such as nano seeds, nanoparticle pesticides, nano ponds, nano foods, and nano packaging have future potential. In the textiles area, nanotechnology future innovation resides in carbon nano fibers, carbon nanoparticles, clay nanoparticles, metal oxide nanoparticles, carbon nanotubes, nanotechnology in textile finishing, and self-assembled nanolayers. Finally, considering the situation of developing countries, nanotechnology may be able to capture the growing future markets of energy, agriculture, water treatment, disease diagnosis and screening, drug delivery systems, food processing and storage, air pollution remediation, construction, health monitoring, disease vector and pest detection control.

Bibliography

J. C. Davies, *Oversight of Next Generation Nanotechnology*, Woodrow Wilson International Center for Scholars, Washington, DC (2009).

C. T. Dervos and P. Vassiliou, Sulfur hexafluoride (SF_6): global environmental effects and toxic byproduct formation. *J. Air Waste Manag. Assoc.*, 2000, **50**, 137.

EPA, Fluorinated Greenhouse Gases. http://www.epa.ie/air/airenforcement/ozone/fluorinatedgreenhousegases/.

M. Levinshtein, S. Rumyantsev, M. Shur, *Properties of Advanced Semiconductor Materials*, John Wiley & Sons, New York, NY (2001).

Nanoscience and Nanotechnologies: Opportunities and Uncertainties, The Royal Society & The Royal Academy of Engineering, Plymouth (2004).

J. Perriére, Rutherford backscattering spectrometry. *Vacuum*, 1987, **37**, 429.

M. Platzer, J. Sargent Jr., U.S. Semiconductor Manufacturing: Industry Trends, Global Competition, Federal Policy, Congressional Research Service, 7-5700, 2016.

B. Ruben in Trace *Analysis of Semiconductor Materials*, Eds. J. Paul Cali R. Belcher, and L. Gordon. Pergamon Press, Oxford (1964).

United Nations Framework Convention on Climate Change: Kyoto Protocol, http://unfccc.int/kyoto_protocol/items/2830.php.

P. Williams, Secondary ion mass spectrometry. *Ann. Rev. Mater. Sci.*, 1985, **15**, 517.

Y. Yoneda, H. Kawazoe and T. Kanazawa, EPR of Gd^{3+} in ZnF_2-BaF_2-RF_3 (R=Rare earth) glasses. *J. Non-Cryst. Solids*, 1983, **56**, 33.

Chapter 8: Forensic Science

Katharyn Hernandez, Michelle Nguyen, Janett Ordonez,
Steven Schara, Olivia Zhang, Pavan M. V. Raja
and Andrew R. Barron

Introduction

Forensic chemistry applies chemistry to criminal investigations in which complex, accurate, analyses of evidence is crucial. Evidence can come from the crime scene, victims, and subjects, either ante-mortem or postmortem. Samples are taken to a laboratory, or in some urgent situations, temporary field laboratories can be set up. Chemists report their findings to the investigators and, if needed, prepare formal testimony for trial. The work of chemists in forensics has far reaching consequences, as evidence produced in court can be life changing for the victim, the subjects, and their families. More broadly, forensic chemistry has the potential to create change in legislation and in the legal system.

Target materials include:
- body fluids,
- drugs,
- toxins,
- residues from explosives, fires, and gunshots.

At the scene of a crime, it is often difficult to distinguish between bodily fluids. However, even if attempts have been made to clean the crime scene by the perpetrator, trace amounts can still be detected, e.g., in the Kastle-Meyer blood test (Figure 8.1), the hemoglobin-catalyzed reduction of peroxide is facilitated by the oxidation of the pale yellow colored phenolphthalin (Figure 8.2) to the intensely pink colored phenolphthalein (Figure 8.3).

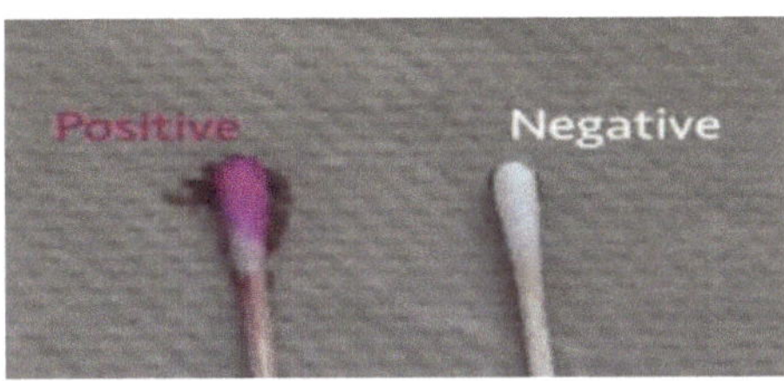

Figure 8.1: The positive detection of blood using the Kastle Meyer test.

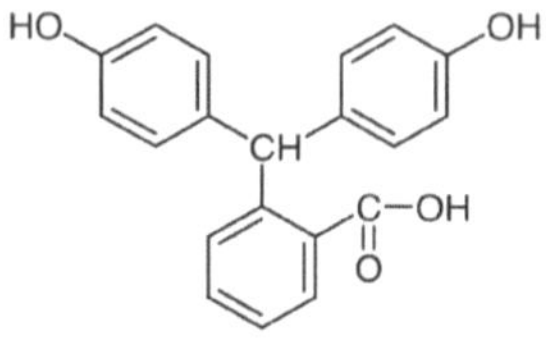

Figure 8.2: Structure of pale yellow colored phenolphthalin (2-(*bis*[4-hydroxy-phenyl]methyl)benzoic acid).

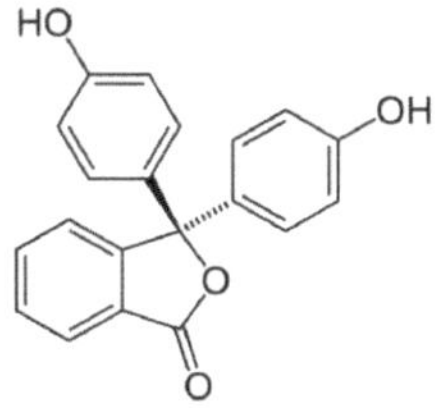

Figure 8.3: Structure of pink phenolphthalein (3,3-*bis*(4-hydroxyphenyl)-2-ben-zofuran-1(3H)-one).

The detection of semen can be enabled with either luminol (Figure 8.4) or fluorescein (Figure 8.5) with the use of an alternate light source (ALS) of 415 nm wavelength. Saliva and urine are other fluids of interest. Once these biological samples have been collected, further extraction of DNA can be performed to uniquely identify subject of interest.

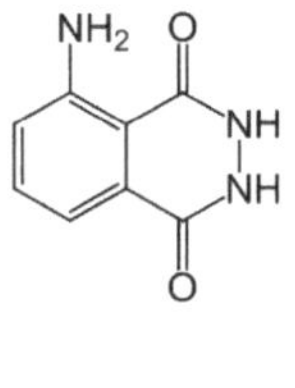

Figure 8.4: Structure of luminol (5-amino-2,3-dihydrophthalazine-1,4-dione).

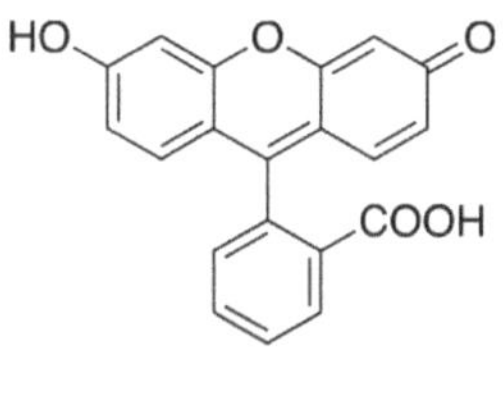

Figure 8.5: Structure of fluorescein (3',6'-dihydroxyspiro[isobenzofuran-1(3H),9'-[9H]xanthen]-3-one).

It is essential to note the physiological condition of persons, which often gives clues to the events that transpired. Alcohol (ethanol), the most common target substance in the field, can be detected via a breathalyzer or through bodily fluids. Illicit drugs, on the other hand, are more difficult, as chemists must distinguish between the metabolites of chemicals from foods and other legal substances and the metabolites of illicit drugs. For example, heroin (Figure 8.6) is metabolized into morphine (Figure 8.7). Tetrahydrocannabinol (THC, Figure 8.8) in cannabis is another common drug and is easily detectable. Pharmaceuticals are important to look for as well as poisons and dangerous chemicals such as lead, arsenic, and mercury.

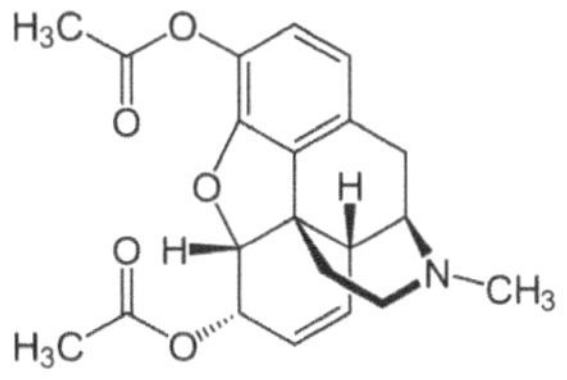

Figure 8.6: Structure of heroin ((5α,6α)-7,8-didehydro-4,5-epoxy-17-methyl-morphinan-3,6-diol diacetate).

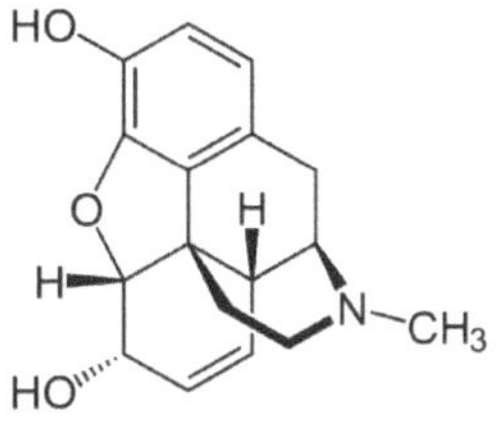

Figure 8.7: Structure of morphine ((4R,4aR,7S,7aR,12bS)-3-methyl-2,4,4a,7,7a,13-hexahydro-1H- 4,12-methanobenzofuro[3,2-e]isoquinoline-7,9-diol).

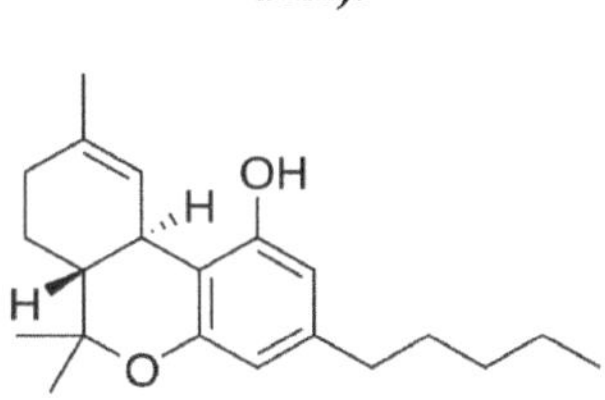

Figure 8.8: Structure of tetrahydrocannabinol (THC, (−)-(6aR,10aR)-6,6,9-tri-methyl-3-pentyl-6a,7,8,10a-tetrahydro-6H-benzo[c]chromen-1-ol).

In residues from gunpowder, chemists search for nitrates and other nitro compounds. After fires, investigators search for accelerants such as gasoline or kerosene that would imply an intentional fire. Furthermore, discovery of C-4 or RDX would indicate a criminal's connection to the military. Other materials of interest include metals from bullet fragments, such as lead, copper, and zinc, as well as paint chips, hair, and other fibers.

Needs and challenges of the industry

With rapid advancements in technology that have greatly aided the work done in forensics, there still remain numerous challenges in the field. Initially established by law enforcement officials, forensics were not built with a strong foundation in science. The lack of a base in research and science is still reflected today within forensic laboratories. Personnel in the field may not be adequately trained in the analytical methods used to characterize evidence. The absence of training and accreditation limits the overall capabilities of forensic labs. Furthermore, forensic labs are constantly backlogged and under-resourced. Part of the taxing work environment can be attributed to the failure to fully educate those who join the facility. Productivity, efficiency, and accuracy of forensic work can very easily be compromised and placed under scrutiny by law enforcement departments.

One of the more striking concerns within forensics is the integrity of protocol and infrastructure across all forensic labs. Even today, labs within the United States fail to enforce standards, quality control, and proper education of forensic scientists. These lax conditions pave the way to contaminated specimens, fraudulent data, erroneous calculations, and lack of documentation. Cumulatively, these major blunders can lead to misinterpretation of evidence and the unjust conviction of an innocent individual. Taking into consideration crime scene investigation methods across the states, officials involved in the general sequence, and the processing of forensic evidence also tend to vary. It is possible to have solely one forensic examiner collect evidence, process evidence, perform the analyses, interpret the evidence, and testify in court. In other regions, it is also feasible to have several investigators collect and process the evidence; scientists conduct the analyses and prepare the conclusive reports. These inconsistencies in policy and protocol, however, leave more room for doubt in the courtroom.

In addition to the organizational and infrastructural issues, there are also inherent analytical challenges that pertain more to the chemistry of the evidence. Analytical chemists, to identify all of their possible metabolites specifically, must study illegal and newly released drugs. The chemical composition of these drugs and their metabolites must be properly studied in order to detect traces of them in future crime scene investigations.

Some of the improvements being worked on in recent years include innovations in toxicology, analysis of trace evidence, and innovations in scale and ease of use of forensic technologies. Beyond always trying to produce instruments, which reduce analysis times, increase instrument sensitivity, and reduce sample preparation, forensics is always trying to find ways to improve identification of drugs or other trace evidence. Since toxicology tests often use gas chromatography or liquid chromatography in tandem with mass spectrometry as a way to identify drugs through database comparison, the field must adapt to test for new drugs as they are created. Additionally, techniques must be adapted to differentiate between substances with the same or similar fragmentation patterns.

Similarly, in trace evidence, there is a necessity of compiling databases to allow for identification of evidence based on statistical data rather than just comparison with another sample, which is currently being done through the creation of explosives databases. Forensics also require ways of analyzing evidence both quickly and accurately, even in the field. To that end, scientists are developing a technology called a lab on a chip, which has been useful in certain medical applications and which will be able to detect DNA, illicit drugs, and explosives residues. Such advances in forensics will no doubt make it a more useful and reliable science in criminal investigation.

Standard analytical protocol

As previously mentioned, one of the biggest issues within the forensic science industry is a failure to establish standard protocols across the nation. While there is no national standard, each individual forensic science lab does establish its flown standard procedures. These standard procedures cover a variety of topics and establish a step-by-step guide for many procedures. Additionally, the guide can cover the uncertainty of measurement, limitations, and documentation pertinent to a specific analysis. Table 8.1 shows an example of the various types of procedures that can be seen in a set of standard operating procedures, as seen in the District of Columbia Department of Forensic Sciences Standard Operating Procedures.

Procedure	Purpose
Forensic biological evidence examination	Procedures used for analyzing evidence for the presence of biological fluids and/or DNA
Phenolphthalein presumptive chemical test for the presence of blood	Procedures used for analyzing evidence for the specific presence of blood
Sodium rhodizonate (Figure 8.9) test	Designed for detecting depositions of vaporous and particulate lead around a suspected bullet hole
Report of results for re-arm examination	Provides a structure for documenting examination results
Acid phosphatase test	Procedures used for analyzing evidence for the specific presence of semen
Modi ed Griess test	Designed to provide specific technique for detecting and preserving patterns of nitrite residues around a suspected bullet hole(s) as a basis for estimating muzzle-to-target distances
Use of alternate light source in stain identification	This procedure is used to help locate possible stains of biological origin for additional testing and identification.

Table 8.1: The standard analytical protocol that many forensic labs follow when conducting analysis of criminal evidence.

Figure 8.9: The structure of sodium rhodizonate.

In an e ort to continue improving the quality of forensic analysis nationwide, the National Forensic Science Improvement Act was signed in 2002. This act provides funding to support education, training, accreditation and certification, facilities, personnel and equipment. Funding from this grant requires that states certify their forensic science laboratories and provide a plan for improving the timeliness and quality of their forensic analysis.

Reporting is yet another aspect of forensic analysis that requires high specificity and transparency. The importance of reporting is emphasized in the courtroom, where the interpretation and presentation of forensic evidence can drastically change a trial's end result. To ensure fairness in a trial, forensic scientists must present the results of all analyses performed on the evidence. Furthermore, forensic scientists must be careful to present their evidence in a scientific, unbiased manner. As seen in the above table, all procedures have their limitations and uncertainty of measurements. These facts must also be disclosed when documenting and when reporting as a key witness.

Key Analytical Techniques

Table 8.2 summarizes the various analytical techniques that are utilized within the field of forensics for analysis of evidence.

Technique	Function of technique	Specific applications
Chemical tests	Identifies presence of specific compounds with simple color and solubility tests	Drugs, toxicology, body fluids, gunshot residue, explosives, paint, documents
Optical microscopy	Magnifies objects to examine evidence	Fibers, paint, drugs, glass, soil, documents, rearms
Transmission electron microscopy (TEM)	Images specimen that cannot be seen through optical microscopy	Paint
Scanning electron microscopy (SEM)	Images surface of samples	Gunshot residue, paint, fibers, glass, documents
X-ray diffraction (XRD)	Determines arrangement of atoms in solids	Explosives, paint, drugs, documents, soil
Infrared spectroscopy (FT-IR)	Identifies functional groups in compound and determines compound by comparing with spectral library	Paint, fibers, polymers, documents, explosives, drugs
Raman spectroscopy	Identifies compound by comparing with spectral library	Drugs, paint, fibers, documents, explosives
UV-visible spectroscopy	Records UV-visible absorbance spectrum of compound	Fibers, paint, documents, drugs, toxicology
Fluorescence spectroscopy (FS)	Records emission spectrum of compound	Body fluids, toxicology, fibers
Continued on next page		

Technique	Function of technique	Specific applications
Nuclear magnetic resonance (NMR) spectroscopy	Identifies pure compound using the interaction of certain nuclei with a magnetic field	Drugs, explosives
Atomic absorption spectrometry (AAS)	Analyzes elements by measuring absorbance and emittance at specific wavelengths of light	Glass, gunshot residue, toxicology
Mass spectrometry (MS)	Characterizes compounds by determining the mass	Drugs, explosives
Thin layer chromatography (TLC)	Separates compounds using liquid organic mobile phase and stationary phase adhered to a glass or plastic plate	Drugs, documents, fibers, explosives
Gas chromatography (GC)	Separates compounds in the vapor phase	Drugs, toxicology, arson residues, explosives
Liquid chromatography (LC)	Separates compounds in the liquid phase	Drugs, toxicology, fibers
Capillary electrophoresis (CE)	Separates charged species with the use of an electric field	Drugs, toxicology, explosives, gunshot residues
Pyrolysis techniques	Produces molecular fragments of compounds at high temperatures that can be identified with GC or MS	Paint, fibers, polymers, documents
Differential scanning calorimetry (DSC) and differential thermal analysis (DTA)	Determines physical and chemical properties of compounds by measuring the change in temperature of the compound as a function of time or temperature	Polymers, fibers
Thermogravimetric analysis (TGA)	Identifies compounds by measuring mass changes as a function of temperature	Polymers, explosives, arson residues
Solid phase extraction	Separates dissolved species from the solvent	Drugs, toxicology
Solvent extraction	Separates compounds based on relative solubility in two different liquids	Drugs, toxicology

Table 8.2: Analytical techniques utilized within the field of forensics for analysis of evidence.

A vast spectrum of analytical techniques is utilized within the field of forensics. With only traces of evidence found at a crime scene, it is then crucial for scientists to carefully and thoroughly examines the samples that are available to them. Ranging from bodily fluids to paint, fibers, and explosive residue, evidence can be analyzed to uncover more information relating to the incident. Bodily fluids commonly characterized using chromatography and spectrometry can provide insight on the toxicology and drug levels of the individual. Chromatography methods, such as GC or HPLC, allow for a separation of chemical components within forensic samples. If performed in conjunction with mass spectrometry, both of these analytical methods can help reveal the toxins, drugs, and other chemicals involved in the crime. The different methods specific to chromatography and spectrometry are often explored in forensics to evade destruction of the sample or to optimize characterization.

In addition to bodily fluids, there are inorganic elements involved in a crime scene. Fibers, glass, inks, explosive residues, and several other forms of evidence can be characterized with techniques such as spectroscopy or microscopy. Implementing these analytical methods based on radiation (IR, UV-visible, Raman, SEM), scientists can acquire more qualitative and quantitative traits of the evidence. Specifically, with microscopy techniques, morphology, layering, and structure of the samples can be determined. These characteristics can be significant when making comparisons to suspect and victim evidence. There are a multitude of other techniques that scientists can use to analyze and characterize evidence. Properly selecting a specific analytical technique is essential and dependent on the quantity, quality, and stable phase of the samples themselves.

Conclusions and future directions

Forensic analysis plays a pivotal role in the world of criminal investigation. There is no doubt that forensics will continue to be an important topic in convicting felons. However, the lack of adequate training for forensic technicians, inconsistency in the protocol used, and understaffed/inadequately supported labs leave a lot of room for improvement. To produce results that are sufficiently unbiased and uncontaminated in the court of law, evidence analysis has to become more regulated and less reliant on techniques like fingerprint analysis, which are subject to inherent bias against a suspect. The necessity of the field to produce accurate scientific results, which are beyond reproach in a court environment, is what makes the chemical analytical techniques used in the field so important. Testing for alcohol, illicit drugs, explosives, gunshot residue, etc. makes the field high stakes in terms of producing results, especially when techniques can run into pitfalls like bias and lack of standards across forensic labs. Thus, it is important that forensics be regulated through legislation like the National Forensic Science Improvement Act. As a science, forensic analysis continues to advance and improve to allow for better evidence analysis in the world of solving crimes.

Increased reliance on analytical techniques like Raman spectroscopy and gas chromatography, which are less prone to biases than other forms of evidence analysis, is one step the field can take to improve the accuracy of results. However, this will likely come at the cost of further increasing backlogs of forensic labs, which already find themselves with too much work and too few resources. Not to mention that as the number of things, which can be analyzed through forensic analytical techniques, increases (more kinds of drugs, explosives, smaller amounts of material) forensic labs will find their services increasingly more in demand. This makes it more di cult to do this work properly. Analytical techniques must adapt to perform investigations at faster speeds without loss of accuracy. As technology advances with innovations like the lab on a chip approach being developed and tested for use in forensic analysis, the results of forensic investigations will no doubt be faster and more reliable than ever.

Bibliography

A. J. Bertino and P. Bertino, *Forensic Science: Fundamentals and Investigations*, 2nd edn.; South-Western: Mason (2011).

R. T. Bowen, *Ethics and the Practice of Forensic Science*; CRC Press, Boca Raton, FL (2018).

S. D. Brandt and P. V. Kavanagh, Addressing the challenges in forensic drug chemistry. *Drug Test. Anal.*, 2017, **9**, 342.

J. M. Butler, U.S. initiatives to strengthen forensic science & international standards in forensic DNA. *Forensic Sci. Int. Genet.*, 2015, **18**, 4.

Forensic Science Standards, https://www.astm.org/Standards/forensic-science-standards.html.

R. M. Gardner and D. Krouskup, *Practical Crime Scene Processing and Investigation*; CRC Press, Boca Raton, FL (2019).

Guidelines for Forensic Scientists, http://www.academyofexperts.org/guidance/guidelines-forensic-scientists.

M. M. Houck and J. A. Siegel, *Fundamentals of Forensic Science*, Elsevier/Academic Press, Amsterdam (2015).

S. Jickells and A. Negrusz, *Clarke's Analytical Forensic Toxicology*, Pharmaceutical Press: London (2008).

N. G. S. Mogollón, C. D. Quiroz-Moreno, P. S. Prata, J. R. D. Almeida, A. S. Cevallos, R. Torres-Guiérrez, and F. Augusto, New advances in toxicological forensic analysis using mass spectrometry techniques. *J. Anal. Methods Chem.*, 2018, **2018**, 4142527.

National Criminal Justice Reference Service https://www.ncjrs.gov/spotlight/forensic/legislation.html.

A. N. Nelwamondo, L. P. Colletti, R. E. Lindvall, A. Vesterlund, N. Xu, A. H. J. Tan, G. R. Eppich, V. D. Genetti, B. L. Kokwane, P. Lagerkvist, B. K. Pong, H. Ramebäck, L. Tandon, G. Rasmussen, Z. Varga, and M. Wallenius, Uranium assay and trace element analysis of the fourth collaborative material exercise samples by the modified Davies-Gray method and the ICP-MS/OES techniques. *J. Radioanal. Nucl. Chem.*, 2018, **315**, 379.

J. H. Ryu, N. Y. Kim, B. W. Kwon, S. K. Suk, J. H. Park, and J. H. Park, Analysis of a third-party application for mobile forensic investigation. *JIPS*, 2018, **14**, 680.

R. Saferstein, *Forensic Science: From the Crime Scene to the Crime Lab*, 3rd edn.; Prentice Hall: Boston (2015).

R. M. Smith, *Handbook of Analytical Separations*, 2nd edn., Elsevier, Amsterdam (2000).

B. Stuart, *Forensic Analytical Techniques*. John Wiley & Sons, Ltd., Chichester (2013).

R. Tino, K. Vizarova, J. Provaznikova, S. Suty, and S. Kirschnerova, Utilization of statistical analysis of FT-IR spectra in forensic examination of paper. *Chem. Pap.*, 2018, **72**, 2265.

US Dept of Justice Committee on Identifying the Needs of the Forensic Sciences Community, National Research Council. Strengthening forensic science in the united states: a path forward; (2012), https://www.nap.edu/download/12589.

A. Widener, Biosecurity: an evolving challenge. *Chem. Eng. News*, 2012, 90, 30.

Chapter 9: Detection of Explosives

Claire Jahnke, Anna Reed, Alice Zhu, Pavan M. V. Raja
and Andrew R. Barron

Introduction

Explosive detection is defined as a non-destructive detection process used to determine if an area, product, or container contains any explosive material. Explosive material is anything that can be initiated to undergo very rapid decomposition resulting in the formation of more stable material, as well as the liberation of heat of the development of the sudden pressure effect. There are many different types of explosive detection that can be used in different fields. Explosive detection is extremely useful for many government agencies, and specifically for homeland security and other counter-terrorism organizations. Unfortunately, in the late 20th century there were many terrorist threats. Examples of these include threats on airplanes and to cities worldwide. The increase in threats led to new developments in the field of explosive detection.

Explosives are a large threat to public safety, and therefore a very important field of research. The effects of bettering detection techniques save lives. The US Bomb Arson Tracking System (BATS) released an explosive incident report in 2015 that details the statistics on explosive incidents. Explosive incidents are any bombings, accidental explosions or other undetermined explosion incidents. In 2015 there were 630 explosion incidents that went unstopped, however there were 6,727 recoveries by BATS. This begins to show the large scope and importance of these techniques.

Explosive detection tools are used in many large public spaces. For example; airports, government offices, industries, churches, bus and train stations, ports, and border control. They are used in almost any space that needs a security entrance. Airports use traditional methods of X-ray and metal detectors. However, these have proven to not be useful for some explosives, so they are in need of new more cutting-edge detection techniques. Explosive threats are constantly evolving, so the research in this field must keep up with the new threats. Worryingly, adversaries continually develop new materials and delivery methods, so scientists must stay on top of explosive research.

Key chemicals and materials

The main materials needed for explosive detection are the samples themselves. What these analytical methods need to be able to do is determine the composition of something and indicate whether or not it is a threat. There are a wide variety of combustible chemicals and materials used in modern explosives, and this clandestine field is continually growing. It is essential on the part of law enforcement and analytical chemists to be constantly evolving their techniques to match the new developments in improvised explosive devices (IEDs).

Although there are many new chemicals being used for explosives, there still need to be methods to detect more classic and well-known chemicals. Such compounds include black powder, a mix of sulfur, nitrates, and charcoal, and trinitrotoluene (TNT, Figure 9.1). While less popular, these compounds and their analogues present just as much of a threat to society. Other common explosive materials to be detected include nitroglycerin (Figure 9.2), perchlorate derivatives, hydrogen peroxide derivatives, and solutions containing ammonium nitrate. All of these compounds and more are extremely volatile and/or reactive and are powerful detonators.

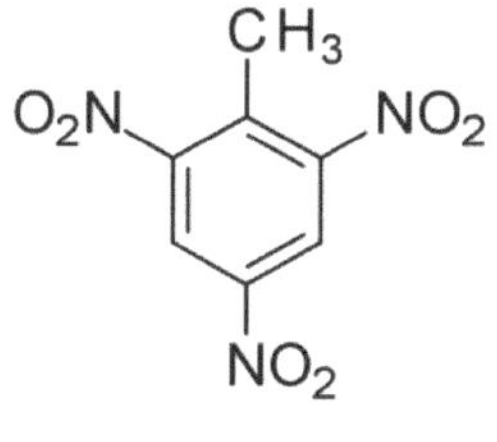

Figure 9.1: Structure of trinitrotoluene (TNT).

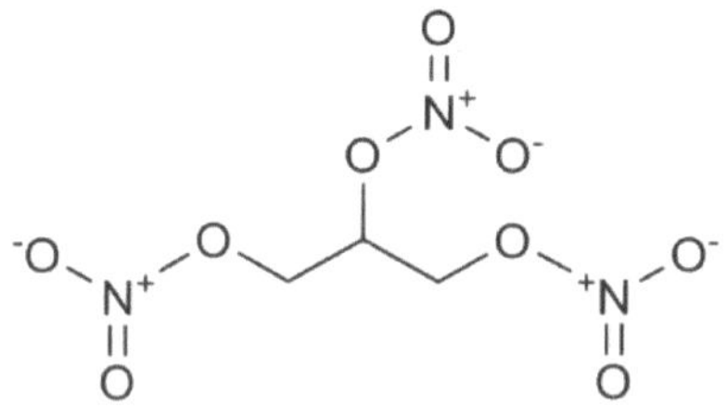

Figure 9.2: Structure of 1,2,3-trinitroxypropane, commonly referred to as nitroglycerin or TNG.

As important as detecting the explosive materials themselves is detecting the precursors to make these improvised units. These precursors could be extremely common items like sugar, fertilizer (ammonium nitrate, NH_4NO_3), and common fuel oil, but when mixed together, in the case of ammonium nitrate and fuel oil (together called ANFO), they can cause blasts like that in the Oklahoma City bombing of 1995. It is crucial that there are ways to identify these compounds separately before detonation, as finding them together in one spot could be an indication of a possible threat.

Key analytical techniques

Table 9.1 details the wide varieties of methods used by laboratories and law enforcement agencies to detect explosives and their precursors. The detection methods listed above t into several categories: spectroscopic, olfactory, general sensory, and nanotechnology techniques. Each category and each method have its unique advantages and disadvantages when it comes to doing this analytical work in the field. For example, for many years spectroscopic methods have been limited to the laboratories because of the size of their instruments. Thus, trained animals, like canines, were the better option for analyzing possible explosives in mobile settings like airports. While the technology has improved for these laboratory methods in recent years, animal detectors are still commonly used and can be very successful for qualitative detection.

In fact, some of the current research focuses on how to mechanize the processes and advantages of canine and animal detection. In canine detection, the animals' more advanced olfactory nerves are capable of detecting traces of explosive material in its odor environment, which the animal is then trained to recognize that specific odorant. One area in which this process has been successfully mechanized is in the electronic noses. Chemical sensors replace an animal's olfactory system and a pattern recognition system tied to an artificial neural network automatically distinguishes what substances are there. These are often more sensitive and accurate than tradition canine methods, with detection limits in the parts per trillion.

Another common qualitative technique involves the use of colormetrics, a technique often utilized with nitroaromatics in general, but specifically with TNT (Figure 9.1). The theory takes advantage of the Janovsky reaction, in which a strong base reacts with the nitroaromatics to produce a color change. This strong base is coated onto a cellulose or glass swab so as to be able to test a specific item in question.

Analytical technique	Purpose
Mass spectrometry (MS)	Identify substance based on mass
Ion-mobility spectrometry	Identify compounds based on mass
Fourier-transform IR spectrometry (FT-IR)	IR radiation to identify functional groups
Terahertz spectrometry	Long-range IR radiation to determine groups
Laser-induced breakdown spectrometry	Vaporize sample to analyze plasma components
Raman spectrometry (Raman)	Attribute vibrations to specific compounds
Animal detectors	Olfactory sensors trained to pick out smells
Electronic nose	Man-made sensor to recognize molecules
Colorimetrics	Color change in presence of nitrogen
Neutron activation analysis (NAA)	Gamma radiation measured and attributed to specific compounds
Chemical sensors	Reactions with sample produce color or conductance change
Electrochemical sensors	Measures electrical output (change in voltage or current) from redox reaction with sample
Enzyme-linked immunosorbent assay (ELISA)	Antibodies specific to explosive bind and fluorescence is measured
Flow-through assay	Elisa assays but with continuous flow
Fluorescence sensors	Either detects sample fluorescence or the quenching of sample on fluorescent material
Molecular-imprinted polymers	Molecules with binding sites for explosive residues
Silicon nanowires	Senses and analyzes sample vapors
Nanotubes	Variety can carry sensors and detect small amounts
Nanoparticles	Can react with sample and cause large scale result

Table 9.1: Common detection methods used in analyzing explosive materials and residues.

These analytical techniques all use different ways to identify potentially dangerous compounds and mixtures, each method has its flown benefits and drawbacks. While some of these methods use properties of the compounds on an atomic level to identify molecules of interest, others, like the chemical and immunochemical sensing methods, rely on the chemical properties of the specific compound when introduced to a reactive chemical or mixture.

More advances in these various technical methods are expected in the near future. There are many fields that are rapidly expanding. For example, the blossoming field of nanotechnology has many applications for explosive detection. There are many ways this area of study can help improve and modify the current detection techniques.

Spectroscopy

One of the more widely used techniques in the lab is spectroscopy, analyzing molecules based on the way they respond to different forms of light. There are many different spectroscopic methods in the overarching field of analytical chemistry, and plenty of them can be applied to explosives detection. Some specific methods include mass spectrometry, ion mobility spectrometry, terahertz spectrometry, Fourier-transform ion spectroscopy, laser-induced breakdown spectrometry, and Raman spectroscopy. While all of these methods have proven useful in identifying and analyzing explosives, some of the above techniques are more commonly used among analytical chemists and law enforcement agents in the field.

Perhaps the most popular spectroscopic method for detecting explosives is ion-mobility spectrometry, which resembles mass spectroscopy in its procedure. In this method, sample particles are exposed to an ionizing source, and sent into a magnetic field. The magnetic environment causes the charged particles to separate by size and take unique amounts of time to travel to the detector. The time-of-flight distinguishes the molecular components of the sample and, using a Faraday plate as a detector, can quickly identify the substance being analyzed and approximately, the concentration of this compound. Research into this detection method has led to the production of many portable machines that have been put into use in airports, and further research is being done into optimizing the process to decrease required sample size further and decrease errors.

Another common use of spectroscopy to explosives detection is Raman spectroscopy, which focuses on the vibrational and rotational changes in molecules when they are exposed to a laser. These changes can be sensed by a detector in the machine, which can then compare these values to known standards in a library. This technique, though not very sensitive at this point in time, is very useful, as it provides results quickly and can be used a distance away from the sample. This is especially of interest for the detection of explosives, as the potential samples being analyzed could be extremely sensitive and dangerous. Research being done into Raman spectroscopy is working on making the system more sensitive and specific, as well as more portable. Should these drawbacks to the Raman method be reduced, this spectroscopy technique could be one of the most effective and ideal ones in the field.

Sensors

Chemical sensors undergo reaction with explosive vapors to result in detectable color or conductivity changes. Examples of chemical sensors include EXPRAY and DROPEX detectors, where clothing/baggage surfaces are wiped with test paper and sprayed with aerosols to detect reactions with explosives.

Electrochemical sensors utilize electron output from redox reactions with explosives at the electrode. Nitroaromatic explosives such as TNT (Figure 9.1) or DNT (Figure 9.3) are ideal for these sensors readily due to their reducible nitro groups. However, these sensors are also very valuable for their ability to detect peroxide explosives, which do not fluoresce, have minimal UV absorption and lack nitro groups, making them generally harder to detect than nitroaromatics.

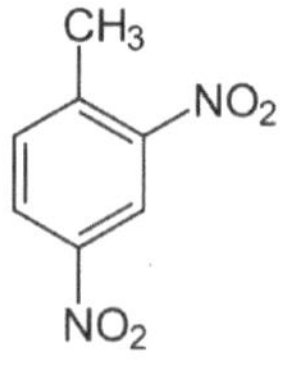

Figure 9.3: Structure of 2,4-dinitrotoluene, commonly referred to as DNT or dinitro, is a precursor to TNT (Figure 9.1).

Immunochemical sensors utilize highly specific antibodies to bind explosives for detection. In competitive ELISA assays, the explosive and explosive analogue conjugated to an enzyme or fluorophore compete to bind the anti-explosive antibody; the resulting fluorescence is measured spectrophotometrically. As concentration of explosive increases, fluorescence decreases as the explosive analogue is outcompeted. This method allows for massive sample throughput with minimal preparation, but there are multiple disadvantages, including multiple wash steps and long incubation times. Flow-through assays circumvent these disadvantages and removes the need for complex lab equipment with continuous fluid flow.

Nanotechnology

Nanoparticle assemblies exhibit a diverse and interesting range of properties that fundamentally differ from the properties of the constituent nanoparticle building blocks. Moreover, they may potentially be used to detect smaller quantities of targeted compounds due to their high surface area to volume ratios. For this reason, nanoparticles have exciting applications in detection fields. Molecular Imprinted polymers are an example of an exciting application. They involve the formation of molecules with specific recognition sites. These sites can be designed to detect specific explosive compounds, for example PETN (Figure 9.4), RDX (Figure 9.5), tetryl (Figure 9.6), and TNT (Figure 9.1).

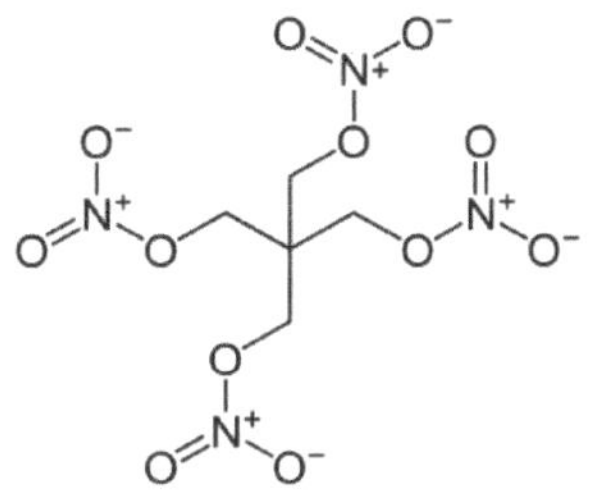

Figure 9.4: Structure of pentaerythritol tetranitrate (PETN).

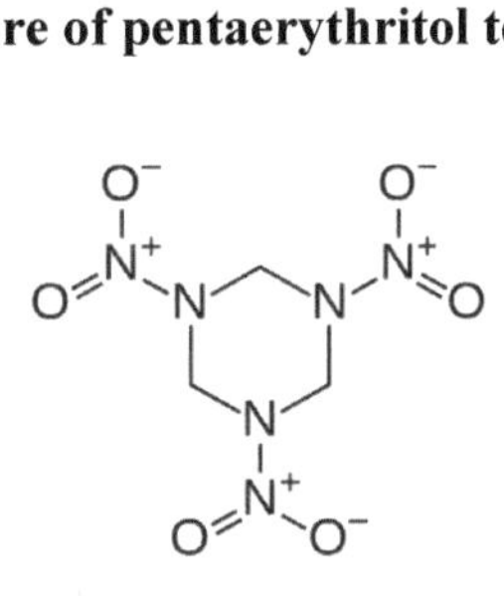

Figure 9.5: Structure of 1,3,5-trinitro-1,3,5-triazinane (RDX).

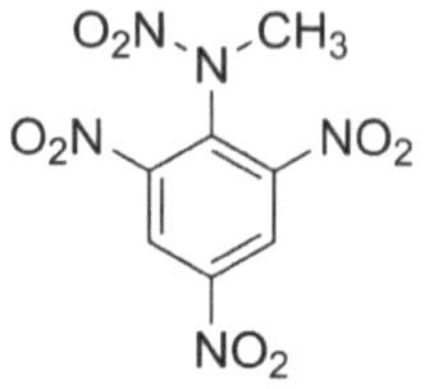

Figure 9.6: Structure of 2,4,6-trinitrophenylmethylnitramine commonly referred to as tetryl.

Silicon nanowires are also a developing technology for explosive detection. They have shown to be extremely useful, and more sensitive than canines. A liquid of vapor containing explosive material is passed over a chip containing silicon nanowire sensing elements. The molecules of the explosive material interact with the surface of the nanowires and as a result induce a measurable change in the electrical properties of the nanowire.

Carbon nanotubes have many applications including the field of explosives. Multi-walled carbon nanotubes (MWCNT) are used to modify glassy carbon electrodes. This increases the surface area of the electrode. Those electrodes are then used to detect TNT, because TNT will accumulate on the surface of the electrode and is aided by the increased surface area due to the MWCNT. Single-walled carbon nanotubes (SWCNT) are also used for explosive detection. They are first aligned in an array on a quartz substrate using chemical vapor deposition (CVD), and then that array is transferred onto textiles. The coated textile then becomes a wearable transmitter. This is a fascinating development that could lead to more wearable chemical sensors.

Nanoparticles can also be used to detect explosive compounds. Gold nanoparticles are used in an aqueous solution. When TNT is introduced to this solution it causes an aggregation of gold nanoparticles which then results in a color change of the solution from red to violet. This an extremely valuable detection tool because it can be observed without further analysis.

Functionalized silver nanoparticles coated on silver molybdate nanowires can also be used to detect TNT through a similar manner as the gold nanoparticles. The TNT complexes the nanoparticles and causes the nanowires to agglomerate together.

Regulations

The Bureau of Alcohol, Tobacco and Firearms (ATF) was formed from over-taking the responsibilities pertaining to alcohol, tobacco, rearms, and explosives from the Internal Revenue Service, in the US. After the terrorist attacks of September 11, 2001, the passage of the 2003 Homeland Security Act added explosives to the name to become the Bureau of Alcohol, Tobacco, Firearms and Explosives, while transferring it from the Department of Treasury to the Department of Justice. Currently, ATF's jurisdiction over explosives includes licensing imports, manufacturing/distribution, permits for shipping/transporting/receiving explosives, and inspection and regulation. ATF enforces both Title XI of the 1970 Organized Crime Control Act and the 2002 Safe Explosives Act. Other departments that have jurisdiction over explosives include Dept. of Transportation over explosives in transit, Mine Safety and Health Administration over explosives at mining sites, Dept. of Defense over explosives in military use, Consumer Product Safety Commission over reworks consumers may buy and Dept. of Homeland Security over high-risk chemicals in industrial/laboratory facilities.

Due to the in-flux of fake explosive detection devices on the market, the Dept. of Justice provides a Guide for the Selection of Commercial Explosives Detection Systems for Law Enforcement Applications. This guide explains how to select and utilize the current explosive detection devices on the market while avoiding bogus devices. Many bogus devices advertise scientific-sounding principles but will be based on superstitions such as dowsing (the act of using rods to find well water) or they will have no backing from regulatory agencies. Currently, the Department of Justice and the scientific community must work in conjunction to enforce regulations on explosive detectors while protecting the public from scientific sounding bogus devices.

Challenges and need for innovation

The explosives detection industry faces the need for continual innovation and product development especially in light of the continual evolution of explosives and improvements to their concealment. Much of the focus in the past has been on the identification and neutralization of nitroaromatic based explosive materials, as these are the most prevalent in traditional explosives. In fact, many explosives are permutations of the same basic compound, cyclonite (Figure 11.20), mixed in different concentrations and with different salts or binding materials. Therefore, as these combinations continue to evolve, so does the need for diverse detection of explosive materials and their precursors.

Some of the most devastating domestic bombings within the United States in recent history have been done not with formal explosives but rather with precursors and crude alternatives. For example, there has been a rise in the use of the ammonium nitrate found in fertilizer and fuel oil (ANFO). Both of these materials are available widely for purchases, and so explosive detection techniques need to evolve to be able to identify when these dangerous precursors are present together. There has also been a rise in non-nitrogen containing explosives, such as peroxide based materials or sarin gas, demonstrating a need for continual innovation.

Another realm of challenges for the explosive detection industry is in the fact that any analytical technique developed must take into account the environment in which they must be used; namely they must be self- contained, mobile, and able to detect from a distance. It is not possible to transport a possible live explosive to a lab for analysis, detection often takes place in busy locations such as transportation hubs, or in remote, unsterile situations if used in military maneuvers. Nor is it safe for security personal to get too close to an unknown object, as it could be set off.

A continual challenge for any detection industry is improving the sensitivity and accuracy of the apparatus and techniques. Any methods of detection used must be powerful enough to detect the presence of dangerous material if it is there, for any missed detection of a wide variety explosives and precursors can have catastrophic consequences. However, it also needs to be discerning and accurate enough to identify the right materials, and not have false positives. An issue with many of the newest technological advances is that while they are much more powerful, they can also give false positives for things like medication or perfumes. At a busy airport checkpoint, for example, this can lead to a waste of valuable time and resources.

Conclusions and future direction

Explosive detection is of the utmost importance in the modern age, and therefore the analytical techniques that ensure our safety must continue to evolve. The techniques developed can be widely applied, whether in military and public defense to private security.

As with any field, there is always a need and room for improvement. In a field such as this, it is pertinent that analytical chemists and bomb technicians both continue to improve their practices and techniques. The future of explosive detection lies in addressing the challenges mentioned above. The ultimate goal of this field is to become more precise and accurate when doing analyses and reduce the number of false readings. Incorrect conclusions drawn from false results can have heavy and disastrous consequences, and the hope is that technological and procedural advances can reduce the likelihood of this happening.

Current research is being done in ways to improve the portability and accuracy of the detection methods described above. Methods like mass spectroscopy and nanoparticle-based sensors are being enhanced by chemists to require less sample material, and less detection time and space. This will greatly improve the quality of results obtained in both the forensic laboratory and the field. The more portable these detection techniques become, the more they can be applied and used to reveal threats. These advancements also work to protect the people using them, by becoming more long-range and less dependent on human operation.

The work being done in explosives detection is a combination of technological advancements and improving procedures so as to be as accurate, quick, and safe as possible. In the politically charged world atmosphere we live in today, the advances in this field have become a key issue in national and global security. While there are many encouraging new advancements, the challenge of developing mobile, accurate, and safe analytical detection methods remains.

Bibliography

A. R. Barron, The chemistry behind explosives and the application of nano technology. *The Detonator*, 2009, **36**, 60.

A. Berliner, M. Lee, Y. Zhang, S. Park, R. Martino, P. A. Rhodes, G.-R. Yi, and S. H. Lim, A patterned colorimetric sensor array for rapid detection of TNT at ppt level. *RCS Adv.*, 2014, **4**, 10672.

Bureau of Alcohol, Tobacco, Firearms, and Explosives, *Explosives*, https://www.atf.gov/explosives.

J. Caygill, F. Davis, and S. Higson, Current trends in explosive detection techniques. *Talanta*, 2012, **88**, 14.

M. Coffey, *Chemical and Explosives Detection*, American Physical Society Task Force on Countering Terrorism, https://www.aps.org/about/governance/task-force/counter-terrorism/coffey.cfm.

Department of Homeland Security: *Explosives Detection and Aviation Screening*, https://www.dhs.gov/science-and-technology/explosives-detection-and-aviation-screening

J. V. Janovsky and L. Erb, Zur kenntniss der directen Brom- und nitrosubstitutionsproducte der azokörper. *Ber.*, 1886, **19**, 2155.

T. Ong, T. Mendum, G. Geurtsen, J. Kelley, Use of mass spectrometric vapor analysis to improve canine explosive detection efficiency. *Anal. Chem.*, 2017, **89**, 6482.

W. J. Peveler, S. B. Jaber, and I. P. Parkin, Nanoparticles in explosives detection – the state-of-the-art and future directions. *Forensic Sci. Med. Pat.*, 2017, 13, 490.

J. Rostberg, *Common Chemicals as Precursors of Improvised Explosive Devices: The Challenges of Defeating Domestic Terrorism*, Master's thesis, Naval Postgraduate School, 2005.

S. Singh, Sensors--an effective approach for the detection of explosives. *J. Hazard. Mater.*, 2007, **144**, 15.

L. Turker, Structurally modified RDX - a DFT study. *Defense Technology*, 2017, **13**, 385.

U.S. Bomb Data Center Explosives Incident Report 2014, https://www.atf.gov/explosives/docs/report/2014-usbdc-expl.

J. Yinon, Handheld chemical-sensing systems come in several varieties and offer advantages over the traditional bomb-sniffing dog. *Anal. Chem.*, 2003, **75**, 99A.

Chapter 10: Outer Space

Jagnoor Benipal, Selin Ergulen, Laura Quinn,
Ricardo Rivera-Maldonado, Pavan M. V. Raja and Andrew R. Barron

Introduction

For thousands of years, humans have looked into space and attempted to analyze and characterize what lies beyond the Earth. The massive technological advancements of the last century have allowed us to undertake these goals with ever-increasing efficiency and accuracy, and today many nations around the world have space agencies dedicated to understanding and exploring the universe around us. These agencies utilize analytical chemistry to serve three main goals: to understand the geology and atmospheric processes of the Earth, to understand better the composition and environments of planets and solar bodies in the solar system and beyond, and to investigate techniques to allow for further human exploration of space.

However, these goals present some significant challenges. Historically, the first challenge that astronomical analysis has faced has been overcoming the distances involved. By utilizing spectroscopic techniques, astronomers can study stars many thousands of light years away. A second major challenge faced when attempting to send humans into space is the risk, especially that associated with the unknown. Gas chromatographs, mass spectrometers, and various kinds of spectroscopic techniques are frequently used to both assess the environments humans could face in outer space and monitor the activities and health of those astronauts currently living in space. The third, and most prohibitive challenge faced by the space industry is that of cost. Sending analytical instruments into space is incredibly expensive, and that expense is directly tied to the weight of the payload. By creating smaller, more efficient analytical instruments, costs of analyzing and exploring outer space can be lowered or mitigated. This article presents an overview of the various techniques associated with analytical chemistry, that are used in space research.

Key chemicals and materials

Although analysis of outer space relies on spectroscopy, many materials and chemicals are necessary. Some, such as tungsten, graphite, and germanium are critical parts of the instruments used in analysis. Tungsten has the highest melting point of any metal, making it an optimal electrode in an X-ray generator as electrons bombard it, causing heat. Other chemicals are necessary for use as reference materials for certain techniques. Overall, many chemicals and materials are crucial for the proper analysis of outer space (Table 10.1).

Material	Applications	Annual global production	Sources of production
Tungsten	X-ray generator for X- ray fluorescence	95,000 metric tons	Midwest Tungsten Service; Xiamen Honglu Tungsten Molybdenum Industry Co., Ltd.; Zigong Cemented Carbide Co., Ltd.
Graphite	Electrothermal ionization in atomic absorption spectroscopy	1.2 million metric tons	CG Thermal LLC; Weaver Industries, Inc.; Industrial Graphite Sales
Lithium-doped germanium semiconductor	Radiochemical neutron activation analysis detector	134,000 kg	Teck Resources Ltd., Yunnan Lincang Xinyuan Germanium Industrial Co., Umicore and Nanjing Germanium Co.
Silica	Fast neutron-activation analysis reference sample for lunar sample analysis	189 million tons	U.S. Silica, Fairmount Santrol, Hi-Crush Partners, Emerge Energy Services
Aluminum oxide	Fast neutron-activation analysis reference sample for lunar sample analysis	130 million metric dry tons	Orbite Technologies Inc., Altech Chemicals Ltd., Baikowski SAS
Cobalt	X-ray source for X-ray diffractometer	100,000 metric tons	Glencore, China Molybdenum, Fleurette Group
Beryl	Standard for diffractometers	6,000 metric tons	Gem fields Company
Mylar (biaxially-oriented polyethylene terephthalate, Figure 10.1)	Window in sample cell containers	4.7 million metric tons	DuPont Teijin Films, Flex Films

Table 10.1: Chemicals and materials used for chemical analysis in outer space.

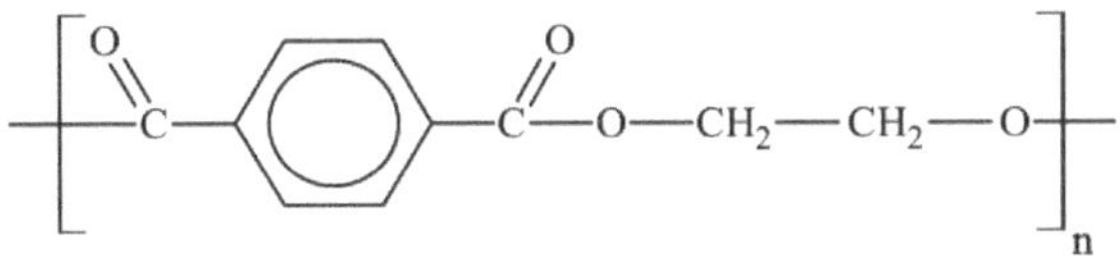

$$\left[\!\!-C(=\!O)\!-\!\!\bigcirc\!\!-\!C(=\!O)\!-\!O\!-\!CH_2\!-\!CH_2\!-\!O\!-\!\right]_n$$

Figure 10.1: Structure of polyethylene terephthalate (PET).

Challenges and innovation needs

The immensity of space presents sizable difficulties, mainly cost and adaptation to a significantly different environment than that of Earth. For centuries, humans have tried gaining knowledge of space and the planets. In recent years, we continue to gain more understanding through modern technology; however, this comes at a massive cost. NASA alone allocated over $200 billion for just a 15-year time period (2005-2020). This incredible sum of money went toward exploration missions, shuttles, maintenance of the International Space Station (ISS), and other expenses. The mysteries and complexities of space have inspired private companies to invest in space discovery as well. NASA reported that there has been over $1.5 billion in public and private expenditure since 2004. Although these amounts already seem unimaginable, these budgets continue to grow with upcoming decades. Further advancements cannot occur without continued financial support.

One of the biggest reasons why cost is so high is because of the drastically different environment that we must adapt to. Space, unlike earth, doesn't have gravity to hold air in place to create a livable atmosphere. Instead, space is almost a perfect vacuum. Additionally, bodies in space are subject to extremely harmful radiation from galactic cosmic rays (GCR) or solar particle events (SPE) such as X-rays, gamma rays, and streams of protons and electrons from solar flares. In order to ensure safety of the people and machines that we send into space, we must account for all of the external factors that could hinder progress.

With all of the foreseeable difficulties, the space industry must constantly consider what innovational needs must be met in space and on Earth. In outer space, for example, the International Space Station (ISS) must be able to withstand not only the subfreezing temperature of space, but also the severe cosmic radiation. In order to protect the station's metal skin from the radiation and keep the astronauts in the station from freezing, scientists have employed Multi Layer Insulation (MLI), Active Thermal Control System (ATCS), and Environmental Control and Life Support System. Most of the ISS is covered with MLI so that the cold cannot penetrate the outside. Because the MLI protects from the cold so efficiently, scientists had to design a system to cool down the internal atmosphere. The ATCS removes heat by a liquid heat-exchange system by circulating the waste heat in cold water and ammonia, and then removing it through honeycomb aluminum panels. Finally, thermal control engineers had to find a way to circulate air flow inside because hot and cold air does not rise and fall in Zero-G conditions. These engineers designed the International Space Station Environmental Control and Life Support System (ECLSS) to work in tandem with the ATCS to control air quality (Figure 10.2).

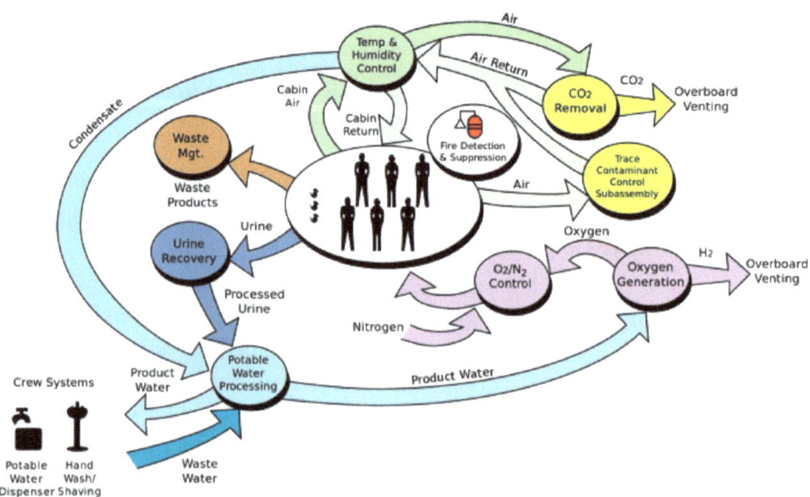

Figure 10.2: The interactions between the components of the ISS Environmental Control and Life Support System (ECLSS).

The space industry needs to be innovative not only in space, but also on Earth. There is potential for discovery to happen anywhere, so technology must be able to handle the demands of any terrain or atmosphere. Engineers, in the pursuit of cosmic knowledge, placed the world's largest radio telescope in Green Bank, West Virginia. This colossal telescope is so sensitive, that it could detect signals as small as a snowflake hitting the ground from 13 billion light years away. To allow for this precision, the town doesn't permit radio, television, or cellphone towers. Interference would wipe out astronomical signals completely. The future of chemistry in space relies on adaptability and optimization of all machines and locations.

Arguably, the most important innovation need for the space industry is the need to continuously advance space technology. In earlier years, emphasis was placed on telescopes and spacecrafts. In 1990, the world's first space telescope, the Hubble telescope, was launched, an e ort that cost over $1.5 billion. Thirty years later, scientists are about to launch the successor to the Hubble telescope. This telescope, the James Webb Space Telescope, cost over $8.7 billion in terms of development and maintenance. However, along this huge price tag, comes observation of larger distances and clearer images. Along with these increased visualization advancements, scientists are working diligently to advance analytical technology, namely through the creation of probes that can withstand extreme environments. Currently, engineers are in the process of launching the Parker solar probe, which would be the first ever visit to a star. This revolutionary machine would give humans insight that was previously unimaginable.

The size, complexities, and mysteries of outer space make further exploration and discovery di cult for the industry. The environment, in space and on Earth, needs to be thoroughly surveyed to accommodate the inconceivably expensive technology that will be used to obtain data. Looking forward, however, the future of the chemistry in space is bright. Opportunities for discovery are exciting and infinite.

Regulations

Due to outer space's enormity, it can't truly be regulated. Instead, nations across the world unite to enact fair and equal laws pertaining to space. Tension between the US and the Soviet Union during the Cold War continued to build in every facet of life, including space exploration. In 1957, the Space Race began when Russia launched the first artificial satellite, Sputnik, into space. This attempt at exploiting the military potential of space drove countries to begin and further cosmic research. Eventually, foreign nations, even ones pitted against each other in decades-long war, came together to begin solving the mysteries in this frontier.

With continuous technological advancements like the ISS, we see how countries collaborate to ensure further progress in making more sense of the unknown. These nations have created treaties, such as the Outer Space Treaty and the Mood Treaty in 1967 and 1979, respectively, to reinforce the idea of keeping space demilitarized. The Outer Space Treaty declares that no weapons of mass destruction can be in Earth's orbit, and no nation can claim sovereignty or occupancy of any celestial space. The Moon Treaty declared that no celestial body, moon included, could be owned. It also declared that all of the states have an equal right to research these bodies in space, but the resource extraction must be shared equally among the countries. Currently, global projects like the square kilometer array (SKA) are growing due to an increasing number of countries that wish to band together with other nations to further advance astrological knowledge. With the aid and combined brainpower of all of these nations, we can continue uncovering all of space's deep mysteries.

Key analytical techniques

Due to the physical barriers inherent to the study of outer space, analytical chemistry applications rely heavily on spectroscopy. However, some applications, such as the analysis of lunar samples, could use a wider variety of techniques. Silicate analysis, for example, is actually the synthesis of various methods and steps in the attempt to analyze minerals and similar samples. Unfortunately, silicate analysis requires larger sample sizes, which are di cult to obtain. Therefore, scientists have begun using X-ray fluorescence spectroscopy, atomic absorption spectroscopy, and other methods to analyze lunar samples as well. Analysis beyond the moon utilizes various spectroscopic techniques to accurately determine elements, reactions, and movement in outer space (Table 10.2).

Conclusions and future directions

The uses of analytical chemistry in understanding and exploring outer space are many and varied, and analytical chemistry is performed throughout our solar system. Spectroscopic and silicate analysis are performed on Earth, to better understand the composition of planets and stars both in our solar system and outside of it. Hundreds of experiments on the International Space Station utilize analytical chemistry, as do the instruments on rovers and probes like the Mars Science Laboratory and the Juno probe. However, the efficiency of these instruments and experiments is limited by cost and, to some extent, by international cooperation, both of which can dictate the amount of data we can collect on the universe around us.

Instrument/Technique	Facility	Location	Purpose
Silicate analysis	Chemical lab	Earth	Elemental analysis of lunar samples
X-ray fluorescence (XRF)	Portable instrument	Earth	Elemental analysis of lunar samples
Atomic absorption spectroscopy (AAS)	Chemical lab	Earth	Elemental analysis of lunar samples
Telescopes w/spectrometers	Observatories	Earth	Determine spectral lines of a stellar body. Determine the doppler shift of a spectrum to analyze the movement of a stellar body
Alpha magnetic spectrometer	ISS	Earth orbit	Analyze cosmic radiation. Detect evidence of antimatter and positrons. Track cosmic events
ANITA interferometer	ISS	Earth orbit	Monitor trace gases
CheMin X-ray diffractometer	MSL	Mars	Determine mineral abundance in samples
SAM: GS-MS with laser spectrometer	MSL	Mars	Detecting carbon-based compounds
DAN: Dynamic Albedo of neutrons	MSL	Mars	Neutron detector searching for evidence of water
UVS: UV imager and spectrometer	Juno	Jupiter	Image the surface of Jupiter & its moons. Further investigate chemical compositions
JADE and JEDI: plasma and energetic particle detectors	Juno	Jupiter	Searching for plasma and energetic particles

Table 10.2: Instrumental techniques.

The future of analytical chemistry in outer space will hinge on when or if we begin human exploration of the Solar System beyond the Moon. In order to do so, extensive testing will likely need to be completed in order to gauge the safety of such journeys, and spectroscopic and chromatographic analysis will help under- stand not just the environments astronauts explore, but the effect of those environments on the astronauts themselves. Many questions, however, remain to be answered about the possibility of such journeys occurring; not least of which is the cost of human exploration in outer space. By improving analytical instruments to be smaller and lighter, costs can be driven down.

However, the nature of the space industry is yet to be fully decided yet: in the near future, we will have to decide upon both the extent of international co-operation and of the private sector in the space industry.

Bibliography

R. L. Korotev, Lunar geochemistry as told by lunar meteorites. *Chem. Erde-Geochem.*, 2005, **65**, 297.

NASA: *NASA Strategic Plan*, https://www.nasa.gov/sites/default/files/atoms/files/6-nasa_2018_strategic_plan_mar2018_tagged.pdf.

NASA: *A Budgetary Analysis of NASA's New Vision for Space Exploration*, https://www.hq.nasa.gov/offce/hqlibrary/documents/o56716382.pdf.

NASA: *Why Space Radiation Matters*, https://www.nasa.gov/analogs/nsrl/why-space-radiation-matters.

NASA: *Staying Cool on the ISS*, https://science.nasa.gov/science-news/science-at-nasa/2001/ast21mar_1.

NASA: *International Relations in Space*, https://www.jpl.nasa.gov/infographics/infographic.view.php?id=11173.

NASA: *International Space Station Alpha Magnetic Spectrometer*, https://www.nasa.gov/mission_pages/station/research/exp.

NASA: *Analyzing Interferometer for Ambient Air*, https://www.nasa.gov/mission_pages/station/research/experiments/explore.

NASA: *Chemistry and Mineralogy X-ray Diffraction (CheMin)*, https://mars.nasa.gov/msl/mission/instruments/spectrometers/.

NASA: *Juno Spacecraft and Instruments*, https://www.nasa.gov/mission_pages/juno/spacecraft/index.html.

J. G. Stears, J. P. Felmlee, and J. E. Gray, Half-value-layer increase owing to tungsten buildup in the X-ray tube: fact or fiction. *Radiology*, 1986, **160**, 837.

A.-C. Zhang, L. A. Taylor, R.-C. Wang, Q.-L. Li, X.-H. Li, A. D. Patchen, and Y. Liu, Thermal history of Apollo 12 granite and KREEP-rich rock: clues from Pb/Pb ages of zircon in lunar breccia 12013. *Geochim. Cosmochim. Acta*, 2012, **95**, 1.